88 Topics in Current Chemistry
Fortschritte der Chemischen Forschung

Organic Chemistry
Syntheses and Reactivity

Springer-Verlag
Berlin Heidelberg GmbH 1980

This series presents critical reviews of the present position and future trends in modern chemical research. It is addressed to all research and industrial chemists who wish to keep abreast of advances in their subject.

As a rule, contributions are specially commissioned. The editors and publishers will, however, always be pleased to receive suggestions and supplementary information. Papers are accepted for "Topics in Current Chemistry" in English.

ISBN 978-3-662-15410-6 ISBN 978-3-540-38970-5 (eBook)
DOI 10.1007/978-3-540-38970-5

Library of Congress Cataloging in Publication Data. Main entry under title: Organic chemistry, syntheses and reactivity. (Topics in current chemistry ; 88) Contents: Rüchardt, C. Steric effects in free radical chemistry. – Birkhofer, L. and Stuhl, O. Silylated synthons. [etc.] 1. Chemistry Organic – Addresses, essays, lectures. I. Rüchardt, Christoph, 1929– II. Series. QD1.F58. vol. 88. [QD255]540'.8s [547] 79-24429

Originally published by Springer-Verlag Berlin Heidelberg New York in 1980
Softcover reprint of the hardcover 1st edition 1980

2152/3140–543210

Contents

Steric Effects in Free Radical Chemistry

Christoph Rüchardt*

Chemisches Laboratorium der Universität Freiburg, Albertstr. 21, D-7800 Freiburg i. Br., Federal Republic of Germany

Dedicated to Professor H. Pommer on the occasion of his 60th birthday.

Table of Contents

---

* This review is an extended version of an article in „Zeitschrift der Sowjetischen Chemischen Mendelejew-Gesellschaft", April 1979.

I Introduction

Steric effects have been discussed in free radical chemistry ever since the discovery of the first free radical, triphenylmethyl *1* by M. Gomberg in 1900[1]. To what extent is the dissociation of its dimer, which was believed to be hexaphenylethane *2*[3] till 1968[2], determined by electronic stabilization of triphenylmethyl *1*[4] or by steric strain in its dimer?

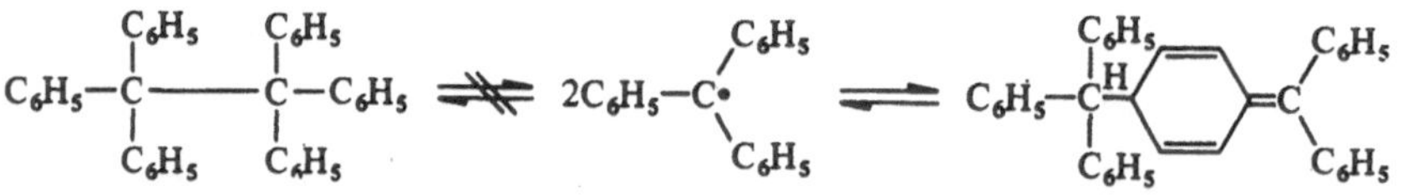

$$(C_6H_5)_3C{-}C(C_6H_5)_3 \;\not\rightleftharpoons\; 2\,(C_6H_5)_2\dot{C}{-}C_6H_5 \;\rightleftharpoons\; (C_6H_5)_3C{-}H\text{-}C_6H_5{=}C(C_6H_5)_2$$

2 *1* *3*

The opinion that stabilization of *1* by resonance was decisive, predominated for a long time and mastered the discussion of the relationship between structure and reactivity in free radical chemistry till quite recently[5]: Accordingly selectivity in free radical reactions was assumed to be mainly due to differences in the thermodynamic stability of the radicals taking part in a reaction or a potential competing reaction.

The recognition[2] that the α, p-dimer *3* is formed in equilibrium with *1* and not the α,α-dimer *2* was interpreted as a result of the smaller steric strain in *3* than in *2*[3]. Also the known strong influence of p-substituents on the equilibrium constants between substituted trityl radicals and their dimers[6] found an obvious explanation in this way. The earlier observation that not only those phenoxy radicals *4* carrying three conjugating phenyl substituents *4* ($R = C_6H_5$)[7a] are persistent[8] but also their

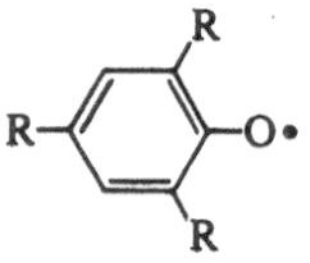

4 (2,4,6-R-substituted phenoxy radical, R–C₆H₂–O•)

t-butylated counterparts *4* ($R = t{-}C_4H_9$)[7b] pointed to the predominating influence of steric effects. Similar results have been obtained in other classes of persistent radicals[7c, 8]. The most convincing evidence for the prime importance of steric effects for the persistence of radicals was provided by the observation of a large series of crowded alkyl radicals like *5–7* over longer periods of time by esr. They do not dimerize for energetic reasons[9, 10].

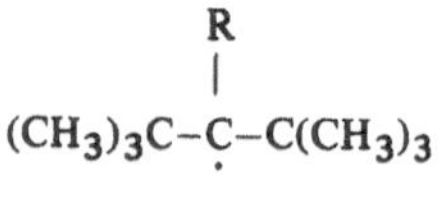

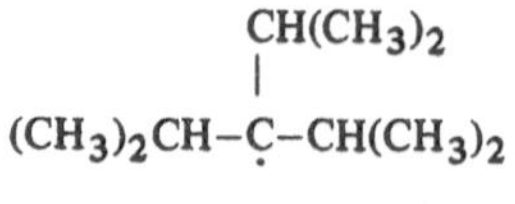

$$(CH_3)_3C{-}\dot{C}(R){-}C(CH_3)_3 \qquad\qquad (CH_3)_2CH{-}\dot{C}(CH(CH_3)_2){-}CH(CH_3)_2$$

5 R = H
6 R = $C(CH_3)_3$

7

Since these developments became known the importance of steric effects on the reactivity of free radical reactions has also been more clearly recognized and more thoroughly investigated[11]. Some more important and more recent results along these lines are the topic of this review.

Finally it has to be remarked briefly that the reactivity and selectivity of free radicals is certainly not only determined by steric and bond energy effects or by the thermodynamic stability of these transients. Polar effects are also important, in particular in those reactions which have "early" transition states e.g., the steps of free radical chain reactions[12]. They are either due to dipole interactions in the ground state or to charge polarization at transition states. FMO-theory apparently offers a more modern interpretation of many of these effects[13].

II Steric Effects in Homolytic Decomposition Reactions

When an alkyl free radical *9* is generated by homolytic cleavage of a C–X bond in its precursor *8*

$$R_3C{-}X \longrightarrow R_3C\cdot + \cdot X$$

8 → *9*

hybridization at the central C-atom changes simultaneously from sp^3 towards sp^2 [14]. All repulsive forces between the substituents R decrease when the bond angles are increased accordingly. Therefore conformational effects can also influence the ease of generation of alkyl radicals.

1 Ring Size Effects

As a model system for demonstrating conformational effects on the rate of radical generation the determination of the influence of the ring size on the rate of formation of cycloalkyl radicals was chosen. Ring size effects on the rate of generation of cycloalkyl carbenium ions were known from the works of Prelog and Brown[15] and were explained by the I-strain[15] i.e., on conformational grounds. During carbenium ion formation the five-ring system loses conformational strain relative to the six-ring system. Cyclopentyl esters therefore solvolyze faster than their cyclohexyl counterparts. Particularly high rate constants were observed for the medium-ring systems. The large transannular nonbonded interactions are partially relieved on ionization due to the formation of planar or nearly planar carbenium ions[16]. When cycloalkyl radicals are generated both effects are also found, in fact the more distinctly, the closer the transition state geometry is approaching the sp^2-state of the radicals[5, 12, 17, 18].

Table 1. Relative rates of formation of cyclic carbenium ions and free radicals from precursors *10–19*[a]

n	*10*	*11*	*12*	*13*	*14*	*15*	*16*	*17*	*18*	*19*
4	2.77	0.03	0.03	0.06	0.297	0.084	0.12	0.23	$1.86 \cdot 10^{-5}$	–
5	124.9	11.5	70.5	2.75	1.18	0.787	0.33	0.47	5.43	0.009
6	≡1.00	≡1.00	≡1.00	≡1.00	≡1.00	≡1.00	≡1.00	1.00	≡1.00	≡1.00
7	108.6	194.0	190	42.8	–	–	1.68	2.27	$2.2 \cdot 10^{4}$	65
8	285.7	1325	–	187	–	–	2.46	4.27	$3.5 \cdot 10^{6}$	>4000
9	44.0	–	–	–	–	–	2.05	4.02	–	–
10	17.8	292	–	–	–	–	1.93	3.26	–	–
11	12.0	–	–	–	–	–	1.89	2.77	–	–
12	–		–	–	–	–	1.76	1.92	–	–

[a] The bonds cleaved in the rate determining step of homolytic decomposition of *11–19* are indicated in the formula.

The five-ring – six-ring effect is larger for the endothermic azo decompositions of *11–13* ($\Delta H^{\ddagger} \approx$ 20–50 kcal/mol)[19–21] than for the decarbonylation of *14* and *15*[22] ($\Delta H^{\ddagger} \approx$ 9–15 kcal/mol)[23]. The five membered cyclic hydrocarbon *18* ($\Delta H^{\ddagger} \approx$ 50 kcal/mol)[24] also decomposes faster than the six membered. The effect is, however, smaller in this example than for the thermolysis of the corresponding azo compounds *12*. This is probably due to the grossly different decomposition temperatures of *18* and *12* and to the overlapping influence of F-strain for *18* (see below). One recognizes from the data in Table 1 that the five-ring – six-ring effect is generally the largest, when α-phenyl- or α-cyano-conjugated radicals are generated. Conjugated radicals require a more strictly planar geometry than unconjugated alkyl radicals[14] (cf. *11–13*). The rate of generation of secondary alkyl radicals from *14* or *17* also responds more strongly to ring size effects than the rate of generation of tertiary radicals from *15* and *16*[25]. The formation of secondary radicals is a more endothermic process. The smallest ring size effect and even an inverse five-ring – six-ring effect is observed in the thermolysis reactions of the peresters *16* and *17*, although all evidence points to a concerted homolytic fragmentation mechanism for these reactions[25]. Apparently, at the transition state of this endothermic reaction the peroxide bond is nearly broken, while the stronger C_{α}-CO-bond is stretched only to a relatively small extent. Therefore, hybridization and geometry at C_{α} have hardly changed. This interpretation is supported by the study of α-CH_3O-[12c], α-CN-[12c] and α-phenyl-substituent effects and by other criteria[5, 12, 18].

Exceptional behavior among the reactions of Table 1 is shown by the thermolysis reaction of *18*. While the direction of the five-ring – six-ring effect is normal, a particular large rate enhancement (10^4–10^6) is found for the thermolysis of the seven and eight membered compounds and an unexpected high thermal stability for the four membered one. Apparently the thermolysis rates of *18* are not only determined by the change in the I-strain but much more by the strong repulsive Van der Waals interactions across the central C-C-bond which are revealed on bond homolysis. A smaller effect of similar nature is recognized in the decomposition rates of *cis*-1-methyl-1-azocycloalkanes *19*[26]. Because of the low activation enthalpies of *cis*-azo decompositions ($\Delta H^{\ddagger} \approx$ 10–15 kcal/mol)[26] the small five-ring – six-ring effect was

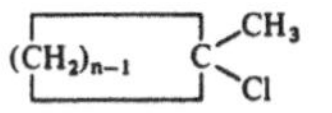

$(CH_2)_{n-1}$ C(CH$_3$)Cl
10[15] 80% EtOH, 25°C

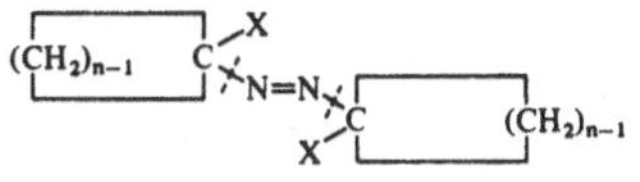

11[19] X = CN, toluene, 80°C
12[20] X = C_6H_5, benzene, 60°C
13[21] X = CH_3, benzene, 200°C

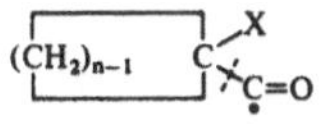

14[22] X = H, 125°C
15[22] X = CH_3, 135°C

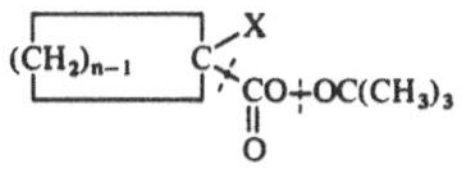

16[25] X = CH_3, benzene, 80°C
17[25] X = H, ethylbenzene, 110°C

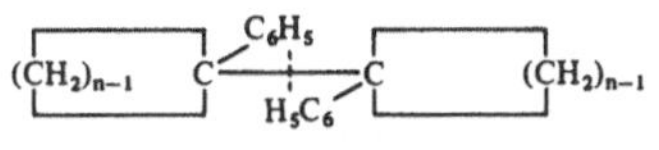

18[24] Octane, 220°C

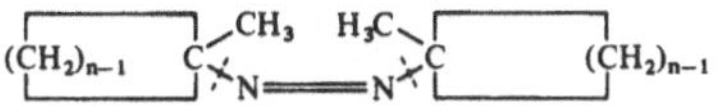

19[26] Ethanol, –28°C

expected because the C-N-bonds are stretched much less at transition state than in the *trans*-azo series. The particularly high rates of thermolysis of *19* (n = 7–8) most probably are due to the release of Van der Waals repulsive interactions between the *cis*-oriented 1-methyl-cycloalkyl groups.

2 Group Size Effects

The influence of the group size on the rate of generation of alkyl radicals has been investigated for the same reactions as mentioned in Table 1 [12a, 27]. Most information is available on the thermolysis of t-azoalkanes *20* (R^1–R^3 = alkyl)[28].

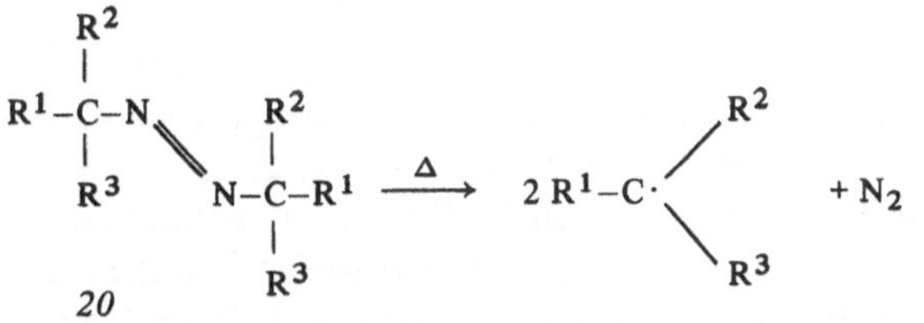

$$R^1R^2R^3C{-}N{=}N{-}CR^1R^2R^3 \xrightarrow{\Delta} 2\,R^1{-}\dot{C}R^2R^3 + N_2$$

20

Qualitatively the same reactivity pattern was observed for the decomposition of *sym.* azonitriles *20* (R^1 = CN, R^2, R^3 = alkyl)[29] and several symmetrically and unsymmetrically substituted azo compounds[30]. A selection of these results is found in Table 2. It is apparent from these data that the thermal stability of *20* decreases as the size of the groups R^1–R^3 increases. Rüchardt et al. have observed that a linear relationship exists between the thermolysis rates of Table 2 and the S_N1-solvolysis rates of corresponding *t*-alkyl-p-nitrobenzoates *21* in 80% acetone-water[28d]. The

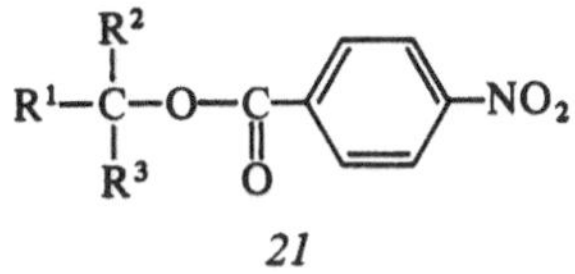

$$R^1R^2R^3C{-}O{-}C(=O){-}C_6H_4{-}NO_2$$

21

Table 2. Rate Constants k_{rel} and activation parameters for the thermolysis of azoalkanes $R^1R^2R^3C{-}N{=})_2$ *20* in hydrocarbon solvents

R^1	R^2	R^3	$k_{rel.}$ (180 °C)[a]	$\Delta H^{\ddagger}$ kcal/mol	$\Delta S^{\ddagger}$ e. u.
CH_3	CH_3	CH_3	≡1.00	43.2[b]	17.7[b]
CH_3	CH_3	C_2H_5	1.19	–	–
CH_3	CH_3	1-C_3H_7	[3.3[c]]	40.7[b]	14.2[b]
CH_3	CH_3	1-C_8H_{17}	2.27	–	–
CH_3	CH_3	2-C_3H_7	3.00	–	–
C_2H_5	C_2H_5	C_2H_5	3.65	–	–
CH_3	CH_3	*t*.But.	5.30	40.9[b]	16.3[b]
			[7.7[c]]		
			[13[d]]		
CH_3	CH_3	*i*-But.	7.51	–	–
CH_3	2-C_3H_7	2-C_3H_7	23.0	–	–
CH_3	C_2H_5	*t*.-But.	36.5	–	–
C_2H_5	C_2H_5	*t*.But.	107	–	–
2-C_3H_7	2-C_3H_7	2-C_3H_7	206	–	
CH_3	CH_3	neo-Pentyl	247	35.6[a]	11.9[a]
			[480[c]]		
			[1320[d]]		
CH_3	2-C_3H_7	neo-Pentyl	453	33.8[a]	9.4[a]
CH_3	CH_3	neophyl	[706[c]]	35.0[b]	11.4[b]
CH_3	neo-Pentyl	neo-Pentyl	[57000[d]]	30.0[e]	5.2[e]

[a] Ref.[28d)] [b] Ref.[28c)] [c] at 150 ° see Ref.[28c)] [d] at 100 °C see Ref.[28a)] [e] Ref.[28a)]

slope of this correlation is approximately 1. Because both series respond in the same way to group size, steric acceleration by relieve of back strain was proposed as common interpretation[28)]. During homolysis of *20* as well as heterolysis of *21* the repulsive Van der Waals interactions between the side chains R^1-R^3 are continuously reduced because the bond angles between these groups are increased during the change of hybridization from sp^3 towards sp^2. Interestingly those examples in Table 2 which carry a neopentyl side chain deviate from the observed correlation. It is assumed that the particularly fast thermolyses rates of neopentyl substituted azo compounds like *22* are due to another type of ground state strain which is releaved on homolysis. It was proposed that due to γ-branching and according to Newman's rule six[31)] heavy Van der Waals repulsions between the methyl hydrogens of the neopentyl groups

H
H
H_3C C
C4 5 6H
H_3C
H_2C_3 2 1N=]$_2$
C
H_3C CH_3

22

and the nitrogen atoms are acting as shown in *22*. The same extraordinary rate enhancing effect of neopentyl side chains was observed for the thermolysis rates of azonitriles *20* (R^1, R^2 = alkyl, R^3 = CN)[29a)] and α-carbomethoxy-azoalkanes *20* (R^1, R^2 = alkyl, R^3 = $COOCH_3$)[32)]. For α-phenyl substituted azoalkanes *20* (R^1, R^2 = alkyl, R^3 = C_6H_5) the relationship between thermal stability and size of the groups R^1 and R^2 is more complex, apparently because the resonance stabilization of the developing radical center at the transition state *23* decreases with increasing group size[33)]. This could be partly due to steric hindrance of resonance[34)]. In addition, however, the transition state *23* is probably reached earlier on the reaction coordinate when the group size of R^1 and R^2 is increased. According to the Hammond principle[17)] this means less C-N-bond stretching and less radical character in *23*. For symmetrical azo compounds *20* (R^1, R^2 = alkyl, R^3 = alkyl, CN, $COOCH_3$, C_6H_5) there is good evidence that both C-N-bonds are cleaved more or less simultaneously in the rate determining step[35)]. This is not generally so for unsymmetrical azo compounds $R^1N_2R^2$ [36)].

[C_6H_5–C(R^1)(R^2)---N≡N---C(R^1)(R^2)–C_6H_5]‡

23

In comparison with the decomposition of *trans*-azoalkanes *20* a much larger group size effect has been found for the thermolysis rates of a few *cis*-azoalkanes *24*. Due to the repulsion of the free electron pairs on the two nitrogen atoms and due to steric interaction between the *cis* oriented alkyl groups *cis* azoalkanes *24* decom-

$R^1R^2R^3$C–N=N–C$R^1R^2R^3$ (*cis*)

24

Table 3. Steric acceleration of thermolysis of *trans*-azoalkanes *20* (180 °C, ethylbenzene) and *cis*-azoalkanes *24* (−28 °C, ethanol)

R^1	R^2	R^3	k_{rel}(*20*)	k_{rel}(*24*)[37)]
CH_3	CH_3	CH_3	≡1.00	≡1.00[a]
CH_3	CH_3	C_2H_5	1.19	4.4
CH_3	CH_3	i-C_3H_7	3.00	64
CH_3	CH_3	i-C_4H_9	7.51	153
CH_3	C_2H_5	C_2H_5	1.87	37
C_2H_5	C_2H_5	C_2H_5	3.65	1428
CH_3	CH_3	t-C_4H_9	5.30	>1600

[a] $k_1 = 0.615 \cdot 10^{-4}\ s^{-1}$

pose at much lower temperatures into radicals[35c, 37]. Although the transition state of this much less endothermic reaction should be located earlier on the reaction coordinate than for the thermolysis of *20*[12a, 17], rates are subject to larger steric acceleration. In addition to the relief of back strain, front strain between the to groups $R^1R^2R^3C$ also becomes important (cf. Table 3).

The rates of homolytic fragmentation of peroxyesters *25* are also enhanced when the size of the side chains R^1-R^3 = alkyl is increased. This is shown for several examples in Table 4. The rate enhancing effect is smaller than for the azoalkane thermolyses

$$R^1-\underset{R^3}{\overset{R^2}{C}}-\underset{O}{\overset{\|}{C}}-O-OC(CH_3)_3 \longrightarrow R^1-\dot{C}\begin{matrix} R^2 \\ R^3 \end{matrix} + CO_2 + \cdot OC(CH_3)_3$$

25

discussed above. Taking into account, however, the multiplicative back strain effect in both alkyl parts of azoalkanes, then the effect of steric acceleration becomes comparable for the thermolysis of *20* and *25*. The different temperature of these two thermolyses reactions may partly be responsible for this. The data of the two series even show a linear correlation with the slope ~ 1 on a logarithmic scale[38b]. Again only the neopentyl substituted compounds deviate from this correlation as discussed previously.

It is somewhat contradictory and not yet fully understood why the back strain effect on the rate of perester decompositions is so large. We had reasoned before from the discussion of conformational effects that the C_α-CO-bond of *25* is only stretched to a small extent at transition state. From an analysis of bond energies[5, 18] it becomes questionable if the homolysis of C-N-bonds (as in *20*) and C-C-bonds (as in *25*) is likely to be directly comparable[5, 12a, 18]. In addition the extent of C_α-CO-cleavage at the transition state of fragmentation of *25* may well be itself dependent on the

Table 4. Steric acceleration of thermolysis of peroxyesters *25* in ethylbenzene at 60 °C[38]

R^1	R^2	R^3	k_1(rel) (60 °C)	$\Delta H^{\ddagger}$ kcal/mol	$\Delta S^{\ddagger}$ e.u.
CH_3	CH_3	CH_3	≡1.00	28.3	5.3
CH_3	CH_3	C_2H_5	1.29		
CH_3	CH_3	1-C_8H_{16}	1.73		
CH_3	CH_3	$(CH_3)_2CHCH_2$	2.30		
C_2H_5	C_2H_5	C_2H_5	3.19		
C_2H_5	C_2H_5	2-C_3H_7	6.50		
CH_3	CH_3	$(CH_3)_3C$	3.4	27.2	4.6
CH_3	CH_3	$(CH_3)_3CCH_2$	2.6	26.5	2.0
2-C_3H_7	2-C_3H_7	2-C_3H_7	32	26.6	6.7

size of the groups R^1-R^3 in *25*. This is indicated e.g., by the small steric acceleration observed when the rates of decomposition of a series of peresters *25* (R^1,R^2 = alkyl, $R^3 = C_6H_5$) with alkyl side chains of different bulk are compared[33].

Table 5. Thermal decomposition of hydrocarbons $R^1R^2R^3C\text{-}CR^1R^2R^3$. Temperature T for $t_{1/2}$ = 1 h, free enthalpy of activation $\Delta G^{\ddagger}$ at 300 °C and strain enthalpy E_S[a]

No.	R^1	R^2	R^3	T[°C] ($t_{1/2}$ = 1 h)	$\Delta G^{\ddagger}$(300 °C) [kcal/mol]	E_S[a] [Kcal/mol]	Ref.
1	CH_3	CH_3	CH_3	490	60.5	7.8	39b, 42)
2	CH_3	CH_3	C_2H_5	420	55.3	14.9	43, 45)
3	CH_3	CH_3	1-C_3H_7	411	53.6	14.8	44)
4	CH_3	CH_3	1-C_4H_9	412	53.9	14.5	44)
5	CH_3	CH_3	i-C_4H_9	384	51.9	18.7	45)
6	CH_3	CH_3	2-C_3H_7	329	46.4	26.3	45)
7	CH_3	CH_3	$(CH_3)_3CCH_2$	321	46.3	27.8	45)
8	CH_3	CH_3	c-C_6H_{11}	315	45.8	32.1	43, 45)
9	C_2H_5	C_2H_5	C_2H_5	285	43.1	42.4	43, 45)
10	CH_3	C_2H_5	c-C_6H_{11}	250	39.6	44.3	43, 45)
11	CH_3	CH_3	*t*-C_4H_9	195	33.7	51.8	45)
12	CH_3	CH_3	H	565	68	2.0	46)
13	C_6H_{11}	C_6H_{11}	H	384	52.1	22.8	45, 47)
14	C_6H_{11}	*t*-C_4H_9	H(D, L)	329	46.7	32.6	45, 48)
15	C_6H_{11}	*t*-C_4H_9	H(meso)	285	42.6	38.5	45, 48)
16	*t*-C_4H_9	*t*-C_4H_9	H	141	29.6	62.7	45, 49)
17	CH_3	H	H	590	69	0	50)
18	H	H	H	695	79	0	51)
19	2.2.4.4 Tetramethylpentane[b]			502	63.9	6.4	45)
20	2.2.3.4.4 Pentamethylpentane[b]			415	55.8	15.1	45)
21	2.2.3.3.4.4 Hexamethylpentane[b]			350	48.8	24.9	45)
22[c]	CH_3	C_6H_5	H	365	50.0	2.8	41)
23[c]	C_2H_5	C_6H_5	H	363	49.7	4.0	41)
24[c]	i-C_3H_7	C_6H_5	H	335	47.4	7.6	41)
25[c]	*t*-C_4H_9	C_6H_5	H	289	42.1	21.4	41)
26[d]	*t*-C_4H_9	C_6H_5	H	303	44.6	18.5	41)
27[c]	*t*-C_5H_4	C_6H_5	H	259	40.3	24.6[e]	41)

[a] Difference in heat of formation as calculated by the force fields according to Ref.[39] (for *1–21*) and Ref.[40] (Set B) for *22–27* and the hypothetical heat of formation of the unstrained molecules[39b, 41].

[b] a statistical correction $k_1 = k_{exp.}/2$ was introduced because this molecule has two equivalent bonds which can be cleaved on thermolysis.

[c] meso-diastereomer

[d] racem.diastereomer

[e] experimental value from heat of combustion

As previously pointed out in the discussion of ring size effects on bond homolyses the largest steric acceleration by bulky substituents is expected for the thermal cleavage of C-C-bonds in tetra- or hexasubstituted ethanes *26*. Im comparison to azoal-

$$\begin{array}{c} R^1R^1 \\ | \;\; | \\ R^2{-}C{-}C{-}R^2 \\ | \;\; | \\ R^3R^3 \end{array} \longrightarrow 2 \begin{array}{c} R^1 \\ | \\ C\cdot \\ / \;\; \backslash \\ R^2 \quad R^3 \end{array}$$

26

kanes the N_2-group separating the two alkyl fragments is missing in *26*. Therefore much stronger front strain interaction across the central C-C-bond is expected in *26* than was found between the alkyl groups in *20* or *24*. This is verified by the results in Table 5. The temperature at which the hydrocarbons recorded in the table decompose with a half time $t_{1/2}$ = 1h varies between 695 °C for ethane and 141 °C for sym. tetra-*t*-butylethane. The difference in free enthalpy of activation is almost 50 kcal/mol in this series! It has been shown that this extremely large rate effect is due to steric acceleration. When the rate constants were correlated with the Taft-Hancock steric substituent constants E_s^c [82] for the halves of the molecules *26* two separate linear correlations were found: one for the compounds *1–11* in Table 5[43] in which the central C-C-bond connects two quaternary centers, the second correlation line is followed by the rate data of a large group of compounds[52] with a central C-C-bond between two tertiary carbons e.g., the compounds *12–16* in Table 5. This separation into two separate correlations is due to differences in structure. The C_t–C_t compounds *12–16* have a gauche ground state conformation which allows for much larger angle deformations in order to escape the building up of ground state strain than anticonformations[47–49].

It was all the more satisfying to find a linear correlation (Fig. 1) between the thermal stability of most aliphatic compounds of Table 5 as expressed by $t_{1/2}$ = 1h or by $\Delta G^{\ddagger}$ (300 °C), and their ground state strain. The strain energies were obtained by force field calculations[39, 40, 51] and confirmed for a selected number of examples by the determination of heats of combustion[48, 49, 52, 53]. This proves that C-C-bond strengths of branched alkanes are mainly influenced by Van der Waals repulsions acting in the ground state of hydrocarbons which are released on bond dissociation. The exponential increase of bond strength for those hydrocarbons *26* with particularly small strain energies (no. 12 and 17–19 in Table 5) is still unexplained[5]. The correlation of Fig. 1 allows the prediction of thermal stabilities of many aliphatic hydrocarbons by force field calculations. It is particularly interesting to note that the diastereomeric compounds no. 14 and 15 of Table 5 have distinctly different stabilities. This was explained on conformational grounds[48]. Another interesting phenomenon is the observation that the slope of the correlation for the aliphatic compounds in Fig. 1 is not –1 but –0.6 as shown by the equation derived from Fig. 1.

$$\Delta G^{\ddagger}\,(300\ ^\circ C) = -0.6\, E_S + 65.6\ \text{kcal/mol.}$$

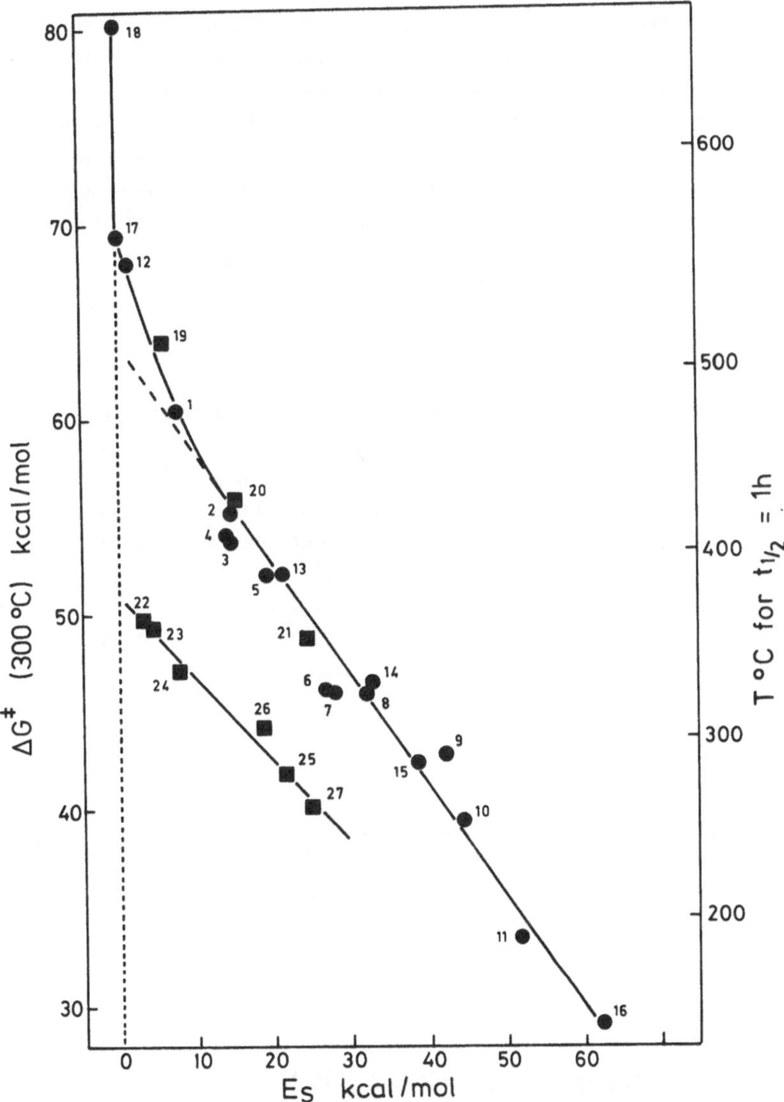

Fig. 1. Correlation between Thermal Stability and Ground State Strain E_S for hydrocarbons *26* (results from Table 5)

This suggests that at the transition state of this homolytic cleavage reaction 40% of the ground state strain is still present. Under the reasonable assumption that the radicals, which are the cleavage products, are more or less strain-free[10b,49], this means, that the recombination of bulky alkyl radicals has an activation barrier of corresponding magnitude. A bond dissociation enthalpy $D_H \sim 76$ kcal/mol is calculated for the C-C-bonds in almost unstrained branched aliphatic hydrocarbons by this correlation in good agreement with the literature value for the central bond of 2.3-dimethyl butane[45].

A corresponding correlation is obtained for the rate constants of α,α'-phenyl substituted alkanes *26* ($R^1 = C_6H_5, R^2 = H, R^3 = $ alkyl) (see Fig. 1)[41]. It has, however, a different slope and a different axis intercept. When both correlations are extrapolated to $E_{Sp} = 0$, a difference of about 16 kcal/mol in $\Delta G^{\ddagger}$ is found. This value is not unexpected because in the decomposition of α,α'-phenyl substituted ethanes (Table 5, no. 22–27) resonance stabilized secondary benzyl radicals are formed. From Fig. 1 therefore a resonance energy of about 8 kcal/mol for a secondary benzyl radical is deduced. This is of the expected order of magnitude[54].

What is the reason for the smaller slope of this correlation?

$$\Delta G^{\ddagger}\ (300\ ^\circ C) = 51 - 0.41\ E_{Sp}\ [\text{kcal/mol}]$$

Two factors are probably contributing: On increasing the strain by increasing the group R^3 in *26* benzyl type radicals are generated which could deviate from planrity and therefore suffer from steric hindrance of resonance[34]. Alternatively, the more strained *26* is, the more the transition state of dissociation of *26* will be shifted in the direction of the hydrocarbon. Its radical character will decrease accordingly and therefore also the size of the resonance effect on the rates[41].

It has to be pointed out, however, that these considerations suffer somewhat from the fact that up to now it was necessary to calculate the strain energies of the phenyl substituted alkanes by a different force field[40] than those of the alkanes[39]

3 Further Steric Effects

When 2-norbornyl type radicals are generated from exo/endo isomeric precursors differences in rate are generally observed. The higher rate of decomposition of the exo-isomer is usually explained on steric grounds[12, 18]. This phenomenon is demonstrated by the following examples:

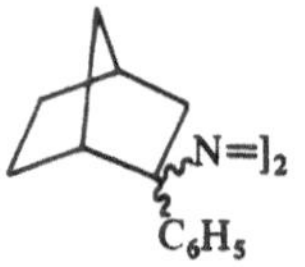

$N=]_2$ / C_6H_5 — $k_{exo(Azo)}/k_{endo} = 116\ (200\ ^\circ C)$[18]

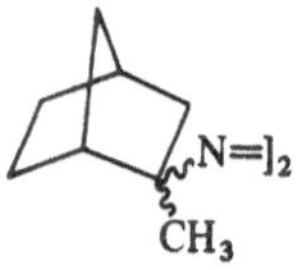

$N=]_2$ / CH_3 — $k_{exo(Azo)}/k_{endo} = 68\ (200\ ^\circ C)$[18]

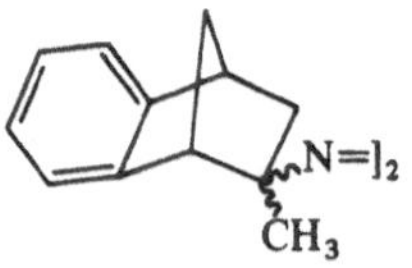

$N=]_2$ / CH_3 — $k_{exo(Azo)}/k_{endo} = 99\ (200\ ^\circ C)$[55]

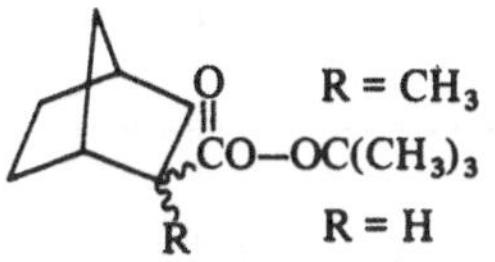

R = CH_3 $k_{exo(CO_2OtBut)}/k_{endo}$ = 6.4 (80 °C)[18, 55]

R = H $k_{exo(CO_2OtBut)}/k_{endo}$ = 4.1 (80 °C)[18, 56]

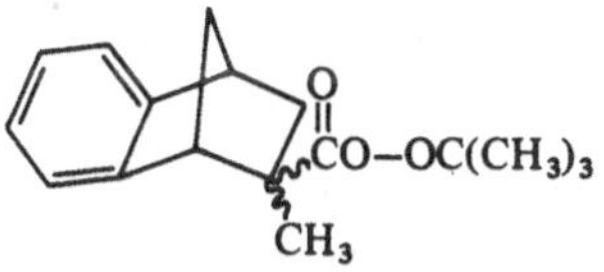

$k_{exo(CO_2OtBut)}/k_{endo}$ = 2.9 (80 °C)[55]

The torsional effect as propsed by Schleyer[57] and steric hindrance of the departing group according to Brown[57] both have been discussed as interpretations of these reactivity series.

The rate of homolytic decomposition of bi- and polycyclic bridgehead azo compounds[18, 58] and peroxyesters[18, 59] decreases with increasing strain of the polycyclic system, because internal ring strain increases further on dissociation. This view is supported by the observation of a linear correlation between the rates of radical generation and the change in strain energy on dissociation as estimated by force field calculations according to Schleyer[58, 59]. For the perester decomposition again a polar effect probably is superimposed.

III Steric Effects in Aliphatic Substitution Reactions

When alkyl radicals take part in atom transfer reactions as acceptors

$$R\cdot + X{-}Y \longrightarrow R{-}X + Y\cdot$$

or as donors

$$R{-}H + Y\cdot \longrightarrow R\cdot + HY,$$

a change in hyridization between the sp^2 and the sp^3 state of the central carbon atom is involved, even though the transition states of these reactions are usually found to be placed early on the reaction coordinates[5, 12, 18]. Because of the different steric interactions of substituents at the central carbon atom and because of the different shielding of the reaction center by these groups in the two hybridization states, steric effects on reactivity are expected in addition to electronic[13] and polar effects[13].

A conformational effect was detected for the H-transfer reactions from cycloalkanes to a series of attacking radicals. The data of Table 6 show that cyclopentane is generally a better H-donor than cyclohexane. The rate ratio is generally largest for the least reactive radicals because the change in hybridization at transition state

Table 6. Relative Rates of H-transfer from cyclopentane (k_5) and cyclohexane (k_6) to radicals X·[60)]

X·	Cl· CCl_4, 0 °C	t-ButO· CCl_4, 0 °C	C_6H_5· CH_3CN, 75 °C	·CCl_3 $BrCCl_3$-CCl_4, 75 °C	Br CH_3CN, 75 °C
k_5/k_6	1.0	1.0	2.8	4.0	3.1

has progressed to the farthest extent in these cases. Increased reactivity is also observed for cycloalkanes of the medium-ring size (C_8–C_{10})[60)].

The well known difference in reactivity in transfer reactions of primary, secondary and tertiary hydrogens is most probably neither due to steric acceleration nor to a difference in electronic stability of primary, secondary and tertiary radicals[5, 12, 18)]. This latter interpretation was favored in the literature until quite recently[5)] because H-transfer reactivity of primary, secondary, and tertiary hydrogens decreases parallel with an increase in the C-H-bond dissociation energy. The suggestion that the drastic change in C-H bond energies is due to a ground state effect[5, 12)] was recently supported by McKean[61)], who observed an interesting correlation between bond dissociation energies and infrared stretching frequencies ν_{CH} for a large group of compounds. Hydrogen bound to carbon atoms which carry good conjugating groups like phenyl or CN deviate distinctly from this correlation. Using the PMO-theory Boldt et al.[13c)] recently recognized fairly good correlations between activation energies of some H-transfer reactions and the superdelocalizabilities $S_r^{(R)}$[13a, b)]. They point out that the principle of maximum overlap, using MINDO-3 data, may serve well in predicting relative rates of H-transfer reactions. Apparently the C-H bond energies are directly related to $S_r^{(R)}$.

Bartlett et al.[62)] on the other hand have found that mainly exo-2-norbornyl halides *28* are obtained from 2-norbornyl radicals *27* and halogen transfer agents. The product ratio of exo-halide *28* and endo-isomer *29* was largest for large halogen transfer agents XY. XY apparently approaches *27* preferentially from the less shielded exo-side. The torsional effect[57)] discussed before is probably also of importance. Similar results were obtained more recently for the transfer of hydroxy groups from peracids to *27*[63)].

+ XY

R

27

X R *28*

R X *29*

Table 7. Relative rates of H-transfer from the 2-position (k_2) and the 1-position (k_1) of adamantane to attacking radicals X· [65, 66]

X·	Cl·	Br·	·CCl_3	$[CH_3-C(=O)-C(=O)-CH_3]^*$	$C_2H_5O-CO-\dot{N}Cl$
k_1/k_2	2–6	9	24	∞	∞

The well known decreased reactivity of hydrogen bound to the bridgehead position of small polycyclic hydrocarbons in transfer reactions is in accord with the steric bridgehead effect[64] discussed above. Although this position usually is shielded to a comparatively low extent an increase in internal strain is expected on bond dissociation. The 2-position of adamantane is more shielded than the 1-position. The ratio of products obtained by radical attack at the 1-position (k_1) and the 2-position (k_2) therefore increases with the size of the attacking radical species[64, 65] as shown by the data of Table 7. Steric hindrance of H-transfer has also been observed in autoxidation reactions. An example is the decreasing reactivity of hydrogens in benzyl position towards attacking cumylperoxy radicals[67] with increasing size of R. In other

$$C_6H_5-C(CH_3)_2-OO\cdot + C_6H_5CH_2R \longrightarrow C_6H_5-C(CH_3)_2-OOH + C_6H_5-\dot{C}HR$$

cases, however, steric acceleration of H-transfer due to relief of back strain was postulated as in the bromination of a series of dibenzo-bicyclooctanes[68]. For example

$\xrightarrow[-HBr]{+Br\cdot}$

the compound with R = CH_3 is more reactive than the corresponding compound with R = H because the steric interaction between *vic.* eclipsed methyl groups decreases in the process of H-transfer. Steric hindrance to H-transfer becomes more pronounced, the bulkier the attacking radical is. This has been favorably used in preparative free radical halogenations for increasing selectivity[69].The best known examples are the chlorinations with N-chloramines in sulfuric acid. Aminium radical cations $R_2NH^{+\cdot}$ are the H-transfer agents in these reactions[70] and their size can be systematically changed by ranging the groups R. An example is the chlorination of isopentane[71]:

$$(CH_3)_2CH{-}CH_2{-}CH_3 \xrightarrow[H_2SO_4,\ 80\%]{R_2\overset{+}{N}HCl} ClCH_2(CH_3)CHC_2H_5 + (CH_3)_2CHCH_2CH_2Cl + (CH_3)_2CH{-}CHCl{-}CH_3 + (CH_3)_2CCl{-}CH_2CH_3$$

R	Relative rate of *prim.*,	*sec.*, and	*tert.* hydrogens[a]
CH_3	0.32	0.93	≡1
$i\text{-}C_3H_7$	0.25	0.70	≡1
neo-C_5H_{11}	0.71	2.70	≡1
t.-C_4H_9	1.70	6.00	≡1

[a] statistically corrected.

In the product determining chain transfer step[70])

$$R_2NH^{+\cdot} + R'H \rightarrow R_2NH_2^+ + R'^{\cdot}$$

of di-*tert.*-butylaminium radical cation the secondary hydrogen is more reactive than the primary and both exceed the reactivity of the tertiary hydrogen, quite in contrast to the usual reactivity order in other hydrogen transfer reactions[5]). For the same reason an unusual product composition of heptylchlorides is obtained in the chlorination of n-heptane using N-chlor-diisobutylamine in H_2SO_4 as chlorinating agent.

$CH_3{-}CH_2{-}CH_2{-}CH_2{-}CH_2{-}CH_2{-}CH_3$

1.3 69.4 22.9 11.3 isomer distribution of chloroheptanes (%)

The high yield of 2-chloroheptane is again due to the least steric shielding of the methylene group in the 2-position from attack by the bulky diisobutyl aminium radical cation. This may be partly due to coiling of the alkane chain in the polar reaction medium. 99% 1-chloroadamantane is obtained by this procedure from adamantane and N-chloro dimethylamine. As the attacking radical has a positive charge this reaction also strongly responds to polar effects. Thus n-alkane derivatives carrying

Product distribution of chlorinations with N-chloro-diisopropylamine

CH_3—	CH_2—	CH_2—	CH_2—	CH_2—	CH_2—X	X
4	85	10	1	–	–	$-O-COCH_3$
6	90	2	2	–	–	$-OH$
4	83	11	1	–	1	$-OCH_3$
7	90	3				$-COOCH_3$

electronegative substituents in the 1-position are chlorinated with high selectivity in the ω-1-position[70]. Similar principles have been used with particular advantage for the selective halogenation of steroids. The bulky reagents

$BrCCl_3$[72, 73] $C_6H_5JCl_2$[73, 74] $C_2H_5{-}O{-}CO{-}NCl_2$[66]

allow in many instances a selective substitution of the most easily accessible 9-α- or 14-α-hydrogens, thus opening synthetic routes to the corticosteroids[74] and cardenolids[73]. Similar selective hydroxylations[75] and fluorinations[76] of steriods have been disclosed which probably also are free radical reactions[74]. A Completely different highly successful approach for selective free radical substitutions in the steriod field based on intramolecular H-transfer was introduced by the Barton Reaction and widely extended since[77]. The key idea was that 1.5-hydrogen transfer[78] is for steric reasons by far the preferred intramolecular mode of transfer. This principle has been elegantly extended in recent years by R. Breslow[73] to "template directed" reactions in which the reagent, e.g., the aryl iodine dichloride moiety, is bound to the steriod substrate *via* alkyl chains of different chain length. A particular advantage has been worked out in the so-called "relay mechanism" in which the substrate bound reagent – e.g., aryliodine dichloride – is generated *in situ* by an external reagent – e.g., SO_2Cl_2 – and a substrate bound precursor of the reagent – e.g., aryliodine[73]. All these reactions will not be discussed in more detail in this review.

An investigation of the competing halogen transfer from $BrCCl_3$ and CCl_4[5, 79] has shown that steric effects are also of importance in atom transfer reactions to alkyl and aryl radicals. Giese[80] investigated very carefully the temperature depen-

$$R\cdot \xrightarrow[k_{Br}]{BrCCl_3} RBr + \cdot CCl_3$$

$$R\cdot \xrightarrow[k_{Cl}]{CCl_4} RCl + \cdot CCl_3$$

$$\kappa = \frac{k_{Br}}{k_{Cl}}$$

pendence of the selectivity κ of this reaction for a large series of alkyl and aryl radicals. Linear Eyring plots of log κ vs. 1/T were obtained over a large temperature range (0 °C–130 °C). Two types of radicals had to be distinguished according to this plot, however, because two separate sheafs of straight correlation lines with intersecting points (isoselective temperature) in the range of 60 ± 20 °C and 50 ± 10 °C, respectively, were observed. The two types of radicals were classified as π-radicals and σ-radicals, respectively[80], but probably a more operational distinction as "flexible" (π) and "nonflexible" (σ) may be preferable[81]. Above and below the isoselective temperature the selectivity series are reversed. There exists therefore no simple structure selectivity relationship. In contrast, however, the scale of differences in activation enthalpies $\Delta H^{\ddagger}_{Cl} - \Delta H^{\ddagger}_{Br}$ of the two competing halogen transfer reactions is independent of temperature. A good linear correlation is obtained, when $\Delta H^{\ddagger}_{Cl} - \Delta H^{\ddagger}_{Br}$ is plotted *vs.* the steric substituent constants E^c_s[82] as shown in Fig. 2 for a series of alkyl radicals.

$$\frac{\Delta\Delta^{\ddagger}_{R}}{\Delta\Delta H^{\ddagger}_{CH_3}} = \delta\, E^c_s$$

With increasing steric shielding of the radical center $\Delta H^{\ddagger}_{Cl} - \Delta H^{\ddagger}_{Br}$ also increases. This steric effect is explained by the difference in X-CCl_3 bond strengths which dominates to a greater extent the selectivity ($\Delta H^{\ddagger}_{Cl} - \Delta H^{\ddagger}_{Br}$) the later the transition state is reached on the reaction coordinate i.e., the more bulky the attacking alkyl radical is.

Because E^c_s-constants for complex groups are obtainable at present only by an empirical procedure[82] a corresponding analysis is not possible for aryl-, vinyl-, and other nonflexible σ-type radicals. This difficulty was overcome recently by the development of a new set of steric substituent parameters $\mathcal{S}_f$ for front strain phenomena. These constants are defined as the difference in heat of formation for the hydrocarbons *30* and *31*. The heats of formation are calculated for this purpose by

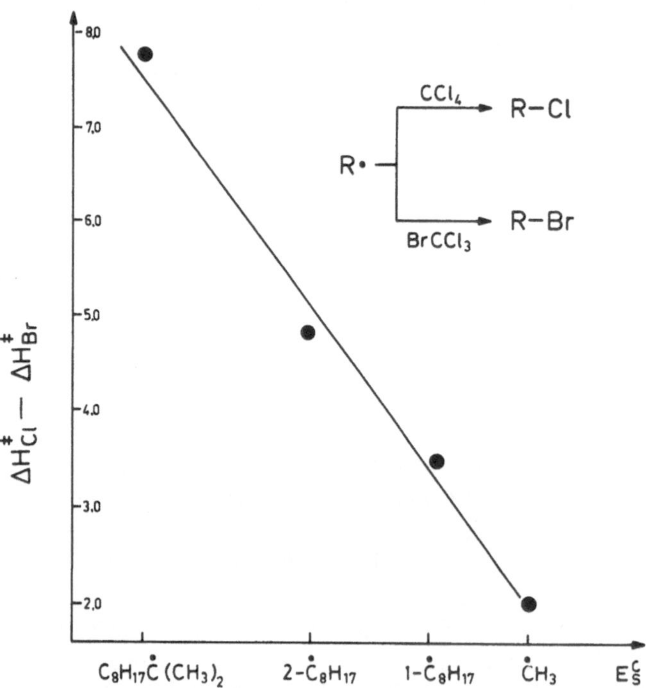

Fig. 2. Linear free enthalpy relationship between the difference in enthalpy of activation for the halogen transfer from CCl_4 and $BrCCl_3$ to alkyl radicals and the steric substituent parameters of alkyl radicals[83]

$R\text{-}C(CH_3)_3$ $R\text{-}CH_3$

30   *31*

molecular mechanics[84]. The plot of Fig. 3 shows that flexible and nonflexible radicals again give two separate correlation lines with $\mathcal{S}_f$ parameters[85]. The nonflexible σ-type radicals have the same geometry as the group R has in the model compounds *30* and *31* used for the computation of $\mathcal{S}_f$. The planar flexible π-type

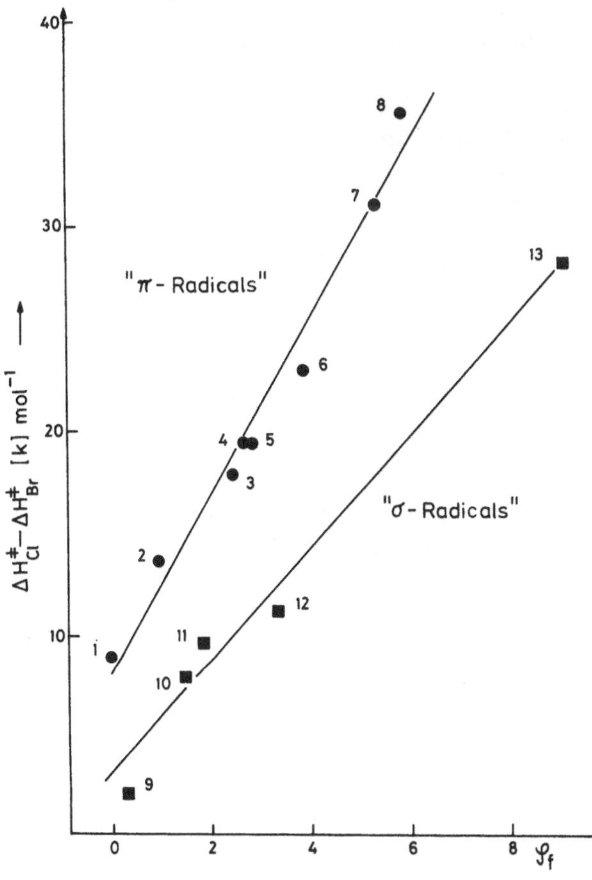

Fig. 3. Correlation of $\Delta H^{\ddagger}_{Cl} - \Delta H^{\ddagger}_{Br}$ [for the halogen transfer from CCl_4 and $BrCCl_3$ to radicals] and the steric substituent constants $\mathcal{S}_f$[84, 85]

No.	Radical	No.	Radical
1	CH_3	8	$CH_3C(C_2H_5)_2$
2	1-C_6H_{13}	9	CH_2=CH
3	c-C_6H_{11}	10	c-C_3H_5
4	2-Bicyclo[2.2.2]octyl	11	C_6H_5
5	2-C_8H_{17}	12	7-Norbornyl
6	$C_4H_9C(CH_3)_2$-CH_2	13	o-t-C_4H_9-C_6H_4
7	$C_8H_{17}C(CH_3)_2$		

radicals on the other hand exert a larger front strain effect towards a reaction partner than predicted from the interaction of its bent analogue structure in *30* and *31*. Therefore the slope of the correlation for the flexible radicals in Fig. 3 is larger than for the nonflexible. This phenomenon supports the assumption of a planar geometry for the flexible π-type alkyl radicals[14]. In addition it stresses the importance of steric effects on free radical substitution reactions. Recently, in a similar analysis for bridgehead free radicals it was shown that $\Delta H^{\ddagger}_{Cl}$-$\Delta H^{\ddagger}_{Br}$ decreases with increasing internal strain of the polycyclic ring systems, although the front strain of these bridgehead radicals increases[86]. The position of the transition states on the reaction coordinate for halogen transfer to bridgehead radicals is apparently mainly determined by the change in i-strain and not so much by f-strain as expected.

Szeimies recently published an impressive example of a steric effect on a S_R2 reaction at carbon for the addition of thiols to the central bond of bicyclo[1.1.0]-systems[87]. From the radical chain addition of thiophenol to *32* the stereoisomeric cyclobutanes *33a* and *33b* are obtained exclusively in 56% yield. The thiylradical

CH_3 CN — C_6H_5SH → H_3C H / NC SC_6H_5 ; H CH_3 / NC SC_6H_5

32 *33a* *33b*

attacks the central C-C-bond in *32* preferentially at the less hindered carbon, generating *34*, although this radical is less stabilized than *35* which would be generated by reversed regioselectivity of thiyl attack on *32*. Steric effects are also known for S_R2-

CH_3 / NC SC_6H_5 ; H_3C SC_6H_5 / CN

34 *35*

substitution at heteroelements. When t.-butyloxy radicals attack di-n-butyl-*t*.-butyl tinchloride a *n*-butyl group is expelled preferentially because it is the smaller ligand which prefers an apical position from which the leaving group usually departs from the addition complex[88].

t–ButO• + $(n$–But$)_2(t$–But$)$SnCl → $[t$–ButO---Sn(t–But)(n–But)(n–But)---Cl$]^{\bullet}$ → $[$Cl---Sn(t–But)(n–But)(Ot–But)---n–But$]^{\bullet}$ → n–But•

IV Steric Effects in Free Radical Addition Reactions

The recognition of anti-Markownikoff orientation when HBr was added to alkenes in the presence of traces of peroxides or air lead to the discovery of the large and important class of free radical addition reactions to unsaturated systems[89]. The anti-Markownikoff orientation of these reactions i.e., the preference of initial radical at-

R(H)C=CH_2 + X· → R–ĊH–CH_2–X
R(H)C=CH_2 + X· ⇸ X–CH(R)–ĊH_2

tack at the less substituted carbon atom of the unsaturated system was interpreted for a long time by the stabilizing influence of α-substituents and in particular of α-alkyl groups at a radical center. As an alternative interpretation, the smaller steric repulsions during radical attack, a double bond at the less substituted end has been discussed[90]. Since the analysis of a large series of bond energies of primary, secondary and tertiary alkyl derivatives had lead to the conclusion that alkyl radicals are not particularly stabilized by α-alkyl substituents[5], the steric interpretation began to enjoy greater popularity[18]. This was supported by the results of intramolecular radical additions leading to cyclization. The homoallylic radical *36* or the 5-hexenyl

·H_2C–H_2C–CH=CH_2 (*36*) ⇸ cyclobutyl radical H_2C–ĊH–CH_2–H_2C ring (*35*)
·H_2C–H_2C–CH=CH_2 (*36*) → cyclopropyl-CH_2· (H_2C–H_2C–CH ring)–ĊH_2 (*37*)

38 (H_2C=CH–CH_2–CH_2–CH_2–ĊH_2) → cyclohexyl radical (*39*)
38 → cyclopentyl-methyl radical, ring CH–ĊH_2 (*40*)

radical *38* cyclize exclusively or at least with high preference to the primary radicals *37* and *40*, respectively, and not or much slower to the secondary radicals *35* or *39*. This was no longer explainable on energetic grounds and a stereoelectronic interpretation was given. During cyclization bond formation by radical attack occurs preferentially at the end of the double bond which is more accessible on steric grounds[78, 91].

Because the addition steps are generally fast and consequently exothermic chain steps, their transition states should occur early on the reaction coordinate and therefore resemble the starting alkene. This was recently confirmed by *ab initio* calculations for the attack at ethylene[92] by methyl radicals and fluorene atoms. The relative stability of the adduct radicals therefore should have little influence on reactivity[12a]. The analysis of reactivity and regioselectivity for radical addition reactions, however, is even more complex, because polar effects seem to have an important influence. It has been known for some time that electronegative radicals X· prefer to react with ordinary alkenes[93] while nucleophilic alkyl or acyl radicals rather attack electron deficient olefins e.g., cyano or carbonyl substituted olefins[94, 95]. The best known example for this behavior is copolymerization[96]. This view was supported by different MO-calculation procedures[92, 97] and in particular by the successful FMO-treatment of the regioselectivity and relative reactivity of additions of radicals to a series of alkenes[13a, 98]. An excellent review of most of the more recent experimental data and their interpretation was published recently by Tedder and Walton[93].

Many examples of the influence of steric effects on reactivity and regioselectivity in free radical additions are known. The anti-Markownikoff regioselectivity apparently is smaller than originally assumed[5, 18, 93, 99] and frequently dependent on the size of the attacking radical[100] as shown by the following data[101]:

$\overset{1}{C}$ $\overset{2}{C}$: $CH_3-CH=CH-R$ — Relative Reactivity for attack at C_1/C_2

R \ X·	$n\text{-}C_4H_9S\cdot$	$t\text{-}C_4H_9S\cdot$
C_2H_5-	1.08	1.10
$2\text{-}C_3H_7$-	1.30	2.55
$t\text{-}C_4H_9$-	1.91	>100:1

Particularly striking is the deactivation of the rate of radical addition by methyl groups at the center of primary attack as shown by the following data[102]:

Relative rates of attack of alkenes by CF_3-radicals

$CH_2=CH_2$	$CH_2=CHCH_3$	$CH_2=C(CH_3)_2$	$CH_3CH=C(CH_3)_2$
↑	↑ ↑	↑ ↑	↑
≡1.0	2.3 0.2	6.0 0.5	2.80

While β-methyl groups excert a slight rate enhancing effect for attack by CF_3-radicals, α-methyl groups reduce it. The small changes in relative rate make it particularly difficult to propose a unique interpretation[93] for these and similar results, because high regioselectivity and low substrate selectivity cannot both be explained on the same energetic grounds e.g., by the different thermodynamic stability of primary, secondary and tertiary radicals[103]. This is even more contrasting for radicals other than CF_3. Ethylene and isobutene have comparable methyl affinities which are 5–10 times higher than those of *cis*- and *trans*-butene[18, 104]. Cyclopropyl radicals attack ethylene even three times faster than isobutene[105]. Similar trends have been observed in detailed investigations of halogenated alkenes[93].

A further interesting example of a steric effect was recently published[106]. The sterically shielded 2.2.6.6-tetramethyl piperidinium radical cation adds to cyclohexene by almost three powers of ten slower than the piperidinium radical cation itself[107].

The combined influences of polar and steric effects and of the strength of the newly formed bond[93] was also recognized in the reaction of α,β-unsaturated carbonyl compounds and similar electron deficient alkenes[95] with organomercurials and $NaBH_4$. For the addition of alkyl radicals to substituted styrenes, ρ assumed a

$$R\cdot + \ >C{=}C(X)< \ \longrightarrow \ R{-}\overset{|}{\underset{|}{C}}{-}\overset{|}{\dot{C}}{-}X$$

$$R{-}\overset{|}{\underset{|}{C}}{-}\overset{|}{\dot{C}}{-}X + RHgH \longrightarrow R{-}\overset{|}{\underset{|}{C}}{-}\overset{|}{\underset{|}{C}}{-}X + R\cdot + Hg^{\circ}$$

$X = -CN;\ -COR;\ -COOCH_3$

small positive value which was, however, dependent on temperature. For the ρ-values of a series of alkyl radicals an isoselective temperature at 90 °C was noted[108]. For the addition of alkyl radicals of different size to maleic anhydride *46* and methylmaleic anhydride *41*, steric effects on the regioselectivity and stereoselectivity became apparent besides polar effects[99, 103]. The regioselectivity series *44:45* is in accord with an explanation by the steric effect in the addition step. The competition constants k_H/k_{CH_3} for the reaction of an alkyl radical with *41* and *46*, respectively, likewise show the influence of a steric effect, but a polar effect as described by the FMO-description could hardly be distinguished. The more nucleophilic attacking radical e.g., t.-butyl, is the more reactive and likewise the more selective[109]. Finally stereoselectivity in the formation of *cis*- and *trans-44* shows that in the second chain step, H-transfer from the less hindered side is prefered, although in this way the less stable *cis-44* is formed in preference to *trans-44.* It has been known for a long time that norbornene is also attacked by radicals from the exo-side[89, 110] with great preference.

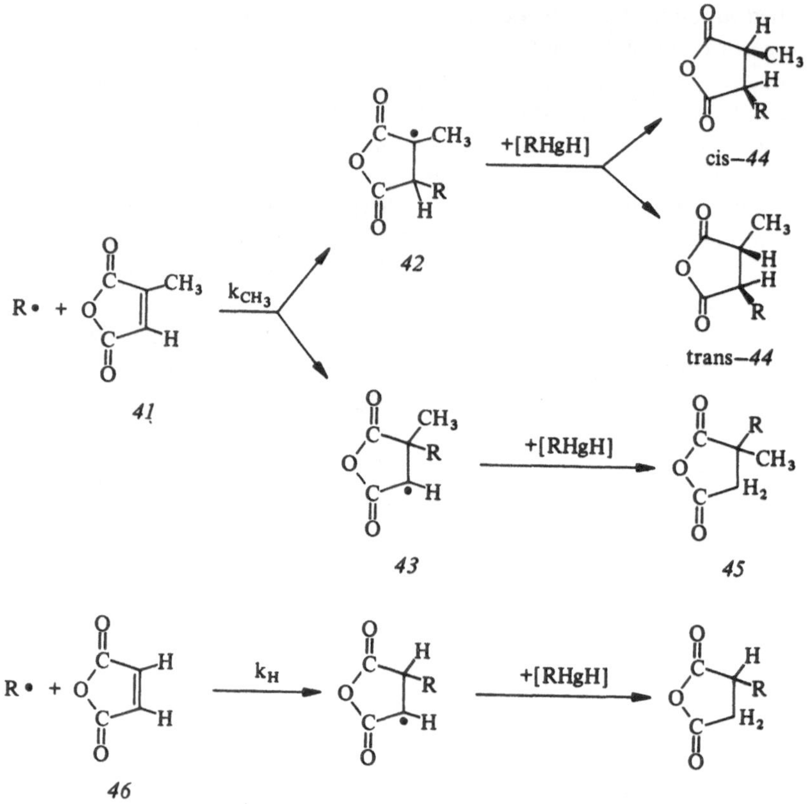

R·	44:45	cis-44: trans-44	k_H/k_{CH_3} (−10 °C)
$(CH_3)_3C\cdot$	99:1	92: 8	13.6
$c\text{-}C_6H_{11}\cdot$	97:3	89:11	9.8
$1\text{-}C_6H_{13}\cdot$	97:3	62:38	6.5
$CH_3\cdot$	98:2	43:57	–

The first step of a free radical aromatic substitution, the formation of the σ-complex, is also an addition step. The o,m,p-product ratio therefore also responds to steric effects. This is shown for the free radical phenylation and dimethylamination of toluene and *t.*-butylbenzene in Table 8. The larger the substituent on the aromatic system and the bulkier the attacking radical, the more p-substitution product is obtained at the expense of o-substitution. In the phenylation reaction the yield of m-product also increases in contrast to the dimethylamination reaction. The substitution pattern of this latter reaction is, in addition to the steric effect, governed heavily by polar effects because a radical cation is the attacking species[113].

Table 8. Steric substituent effects in free radical aromatic substitutions

R·	Toluene			t.-Butylbenzene		
	%o	%m	%p	%o	%m	%p
$C_6H_5\cdot$ [111]	63	21	16	24	49	27
$(CH_3)_2NH^{+}\cdot$ [112]	5.6	22.6	71.8	0	14.6	85.4

Even more pronounced steric effects have been observed for the free radical alkylation of protonated N-heterocyclic bases by the procedure of Minisci[69, b, d]. Quinoline is attacked selectively in the 2- and 4-position by nucleophilic alkyl radicals in sulfuric acid. The largest radicals, *t.*-butyl, react exclusively in the 2-position because of steric hindrance by the peri-hydrogen when attack occurs at the 4-position.

R·	% 2-alkylquinoline	% 4-alkylquinoline
$CH_3\cdot$	23	25[a]
$1\text{-}C_3H_7\cdot$	28	36[a]
$2\text{-}C_3H_7\cdot$	13	26[a]
$t\text{-}C_4H_9\cdot$	100	0

[a] besides 2.4-dinitroquinoline.

Steric effects, although clearly recognized, introduce relatively small rate retardations or increases in selectivity in all these examples, probably because the transition states of all these addition reactions are rather loose ones, i.e., they occur early on the reaction coordinate when the distances between the radical and the substrates are still rather large[92, 93, 97]. An extreme example of a free radical reaction which does not response heavily to steric effects, is the S_{RN}1-substitution reaction of Kornblum[114] by which bonds between two quaternary carbons can be formed with great ease and in good yield, as is shown by one of many published examples[114]. The decisive step

$$O_2N\text{-}C_6H_4\text{-}C(CH_3)_2\text{-}Cl + (CH_3)_2\overset{\ominus}{C}\text{-}NO_2 \longrightarrow O_2N\text{-}C_6H_4\text{-}C(CH_3)_2\text{-}C(CH_3)_2\text{-}NO_2$$

in the chain reaction is the attack of a p-nitrocumyl radical at the carbanion center generating a new aromatic radical anion. The rate of this new type of reaction is apparently extremely high and therefore does not respond strongly to steric effects.

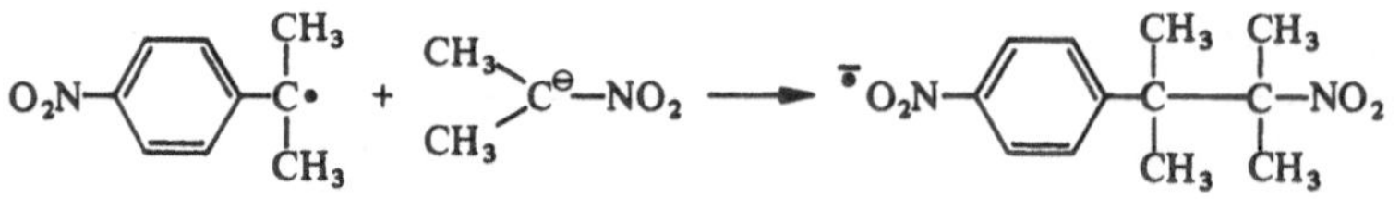

$O_2N-C_6H_4-\dot{C}(CH_3)_2 + (CH_3)_2C^{\ominus}-NO_2 \longrightarrow {}^{\bar{\bullet}}O_2N-C_6H_4-C(CH_3)_2-C(CH_3)_2-NO_2$

In a similar fashion therefore, quaternary substituents can also be introduced to aromatic ring systems by the aromatic counterpart $S_{RN}1$-procedure as investigated mainly by Bunnett[115]. In an extreme situation of steric shielding, however, a response

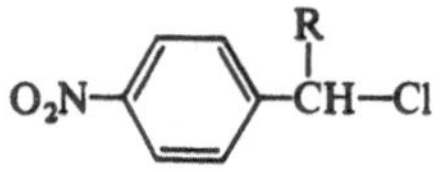

$O_2N-C_6H_4-CHR-Cl$

to steric effects has been detected. 1-alkyl-p-nitrobenzyl chlorides react with the anion of 2-nitropropene with C-alkylation when R = CH_3, C_2H_5, but with 0-alkylation when R = i-C_3H_7 or t.-butyl[116].

V Steric Effects in Dimerization and Disproportionation Reactions

The unusual persistence of many highly branched alkyl radicals[9, 10] mentioned in the introduction proves that radical dimerizations can be hindered or even suppressed by the steric effect of bulky groups. For the dimerization of di-tert-butylmethyl e.g., an activation barrier of about 20 kcal/mol was estimated[49]. Most examples of persistent alkyl radicals, as e.g., *5* and *6*, have no β-hydrogens which are the prerequisite for disproportionation to occur. Triisoproplymethyl *7*, however, is also persistent although β-elimination of hydrogen should lead to destruction of this radical in the course of disproportionation with another radical. It is presumed[9] that *7* has a conformation *47* in which the β-hydrogens are arranged in the nodal plane of the SOMO. Therefore, H-transfer to an attacking radical and formation of a double bond cannot be a synchronous process. Very recently, however, Berndt et al.[10b] have

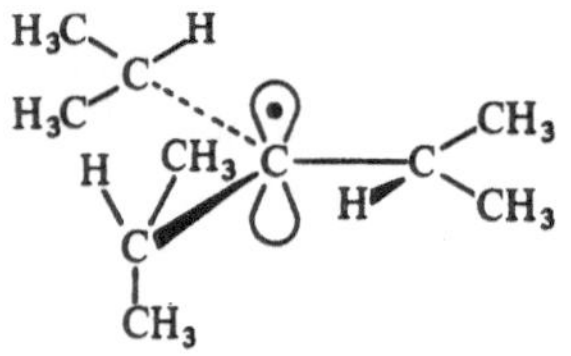

47

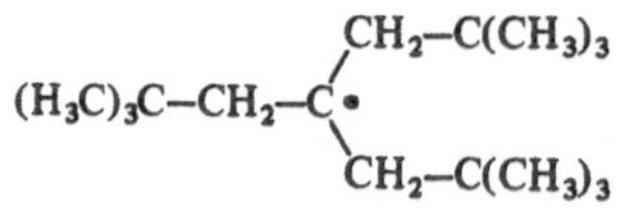

$(H_3C)_3C-CH_2-\dot{C}(CH_2-C(CH_3)_3)_2$

48

reported that trineopentylmethyl *48* and a few other neopentyl substituted methyl radicals also show remarkable persistence. It was not reported whether their decay is a unimolecular or bimolecular process.

In general, the rate ratio of the disproportionation k_d and dimerisation k_c increases with the bulk or size of the radicals concerned[117]. For simple alkyl radicals even a

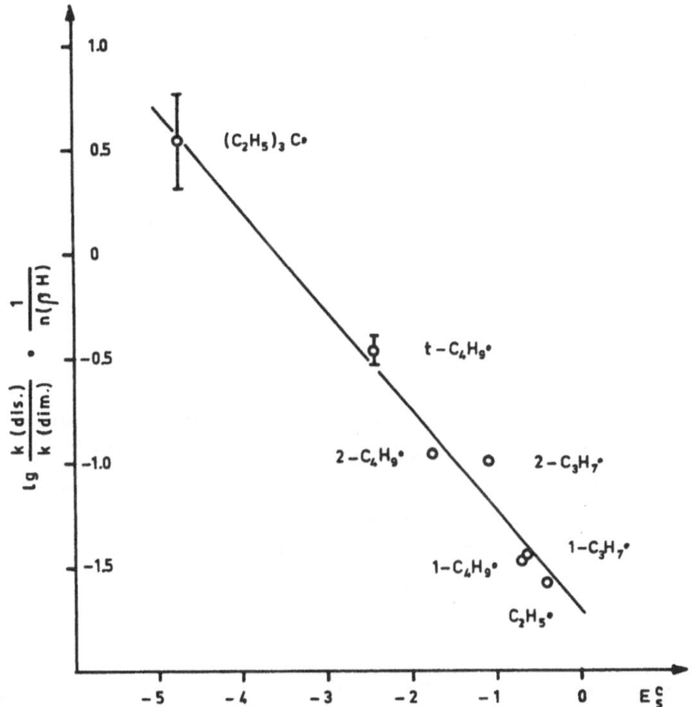

Fig. 4. Relation between the statistical corrected ratio of rates of disproportionation and dimerization of alkyl radicals and their E_s^c-constants[43] $\lg \frac{k_d}{k_c} \cdot \frac{1}{n_{\beta H}} = -0.48\ E_s^c - 1.73\ (r = 0.9901)$ $n_{\beta H}$ = number of β-H-atoms in the radical

linear relation between log k_d/k_c (statistically corrected) and Taft's steric substituent constants E_s^c was found[43] (see Fig. 4). The interpretation of this steric effect is a more more subtile problem than recognized on first sight. Schuh and Fischer[118] have shown by an investigation of the influence of temperature and solvent viscosity on the termination constant, as well as k_d and k_c, for t.-butyl radicals that this effect cannot be explained simply by the greater steric hindrance of approach of the two radicals for dimerization than for disproportionation. The termination constant 2 k_t of the self reaction of *t.*-butyl radicals is diffusion controlled and requires no activation. Observed large solvent and temperature dependences of k_d/k_c were ascribed to anisotropic reorientation motions of the radicals during their encounter in the solvent cage. This may also be the reason for the low probability of recombination of 2-cyano-2-propyl radicals as deduced from CIDNP-experiments[119].

Recently, an interesting example of stereoselective radical dimerization was described which awaits explanation. It was found that radical *49* (X = p-Cl; R = t-butyl) dimerizes diastereoselectively[120] to the more stable D, L-diastereomer in contrast to other radicals *49* with smaller side chains R. It has not been clearly decided so far

X·	R	Yield ratio D, L: meso
H	CH_3	1:1
H	C_2H_5	1:1
Cl	t-C_4H_9	1.66:1

whether the dimerization of this rather bulky radical is an activated process or a diffusion controlled one, and whether diastereoselectivity is due to a difference in free activation enthalpy for the two possible dimerization modes or due to anisotropic orientation motions as discussed by Schuh and Fischer[118]. The temperature dependence of the diastereoselectivity of this dimerization was found to be quite small. The influence of solvent[118] is actively being investigated at present at the author's laboratory.

Acknowledgements. It is a pleasure to thank my coworkers whose names are mentioned in the references and in particular to Dr. Beckhaus for excellent collaboration and important contributions to our own work reported in this review. We are also indepted to the Deutsche Forschungsgemeinschaft, the Fonds der Chemischen Industrie, and BASF AG for financial support of our work.

VI References

1. For a stimulating discussion on historical developments see: a) McBride, J. M.: Tetrahedron *30,* 2009 (1974), b) Walling, C.: Organic Free Radicals. W. A. Pryor (ed.). ACS symposium Series 69, p. 3 (1978)
2. Lankamp, H., Nauta, W. T., McLean, C.: Tetrahedron Lett. *1968,* 249; Staab, H. A., Brettschneider, H., Brunner, H.: Chem. Ber. *103,* 1101 (1970); Volz, H., Lotsch, W., Schnell, H.W.: Tetrahedron *26,* 5343 (1970)
3. Force Field Calculations of *2* were recently published by Hounshell, W. D., et al.: J. Am. Chem. Soc. *99,* 1916 (1977); for an x-ray analysis of the first isolated derivative of *2* see Stein, W., Winter, W., Rieker, A.: Angew. Chem. *90,* 737 (1978); Angew. Chem. Int. Ed. Engl. *17,* 692 (1978). The unusually short central C-C-bond length is in conflict with the low thermal stability of this compound and with the known long C-C-bonds in other crowded hydrocarbons e.g., Destro, R., Pilati, T., Simonetta, M.: J. Am. Chem. Soc. *100,* 6509 (1978)
4. Stolle, F. V. D., Rozantsev, E. G.: Russ. Chem. Rev. *42,* 1011 (1973); Kessler, H., Moosmayer, A., Rieker, A.: Tetrahedron *25,* 287 (1969); Stein, M., Rieker, A.: Tetrahedron Lett. *1975,* 2123
5. Rüchardt, C.: Angew. Chem. *82,* 845 (1970); Angew. Chem. Int. Ed. Engl. *9,* 830 (1970)
6. Ballester, M. in Free Radicals in Solution, p. 123, London: Butterworths 1967; Pure and Appl. Chem. *15,* 123 (1967)

7a. Dimroth, K., Kalk, F., Neubauer, G.: Chem. Ber. *90,* 2058 (1957); b. Müller, E., Ley, K.: Z. Naturforsch. *8b,* 694 (1953); Cook, C. D.: J. Org. Chem. *18,* 261 (1953); c. Buchachenko, A. L., Stable Radicals, New York: Consultants Bureau 1965; Forrester, A. R., Hay, J. M., Thompson, R. H., Organic Chemistry of Stable Free Radicals, New York, London: Academic Press 1968
8. For a discussion concerning the difference between thermodynamic stability of radicals and their kinetic persistence e.g., due to steric effects see Ref.[9)]
9. Griller, D., Ingold, K. U.: Acc. Chem. Res. *9,* 13 (1976)
10. For further examples of persistent nonconjugated crowded radicals see e.g., a. Schreiner, K., Berndt, A.: Tetrahedron Lett. *1973,* 3411; b. Schlüter, K., Berndt, A.: Tetrahedron Lett. *1979,* 929; c. Mendenhall, G. D., Griller, D., Ingold, K. U.: Chem. in Britain *10,* 248 (1974); d. Mendenhall, G. D.: Sci. Prog. Oxford *65,* 1 (1978)
11. For an early pioneer work see Ziegler, K.: Angew. Chem. *61,* 168 (1949)
12a. Rüchardt, C., Mechanismen radikalischer Reaktionen, Forschungsbericht des Landes Nordrhein-Westfalen Nr. 2471, Westdeutscher Verlag, Opladen 1975; Rüchardt, C., Topics Curr. Chem. *6,* 251 (1966); Russ. Translation: Uspekhi Khim. XXXVII, 1402 (1968); b. Davies, W. H., Glenton, J. H., Pryor, W. A.: J. Org. Chem. *42,* 7 (1977), c. Rüchardt, C., Mayer-Ruthardt, J.: Chem. Ber. *104,* 593 (1971); Rüchardt, C., Pantke, R.: Chem. Ber. *106,* 2542 (1973)
13a. Fleming, Ian, Frontier Orbitals and Organic Chemical Reactions p. 182, London: Wiley 1976; b. Fukui, K., Theory of Orientation and Stereoselection, p. 47ff. Heidelberg-New York: Springer 1975; c. Bartels, H., Eichel, W., Riemenschneider, K., Boldt, P.: J. Am. Chem. Soc. *100,* 7740 (1978); d. Giese, B. und Meixner, J., Angew. Chem. *91,* 167 (1979); Angew. Chem. Int. Ed. Engl. *18,* 154 (1979)
14. Although many spectroscopic and chemical investigations of alkyl free radicals have been interpreted by a planar geometry, several more recent results point to a slightly pyramidal arrangement of the bonds at the central C-atom of some alkyl radicals. c.f. Kaplan L., Free Radicals, Kochi, J. K., (ed.), Vol. 2, p. 361, 1st edition, New York: Wiley 1973; Fort, R. C., Schleyer, P. v. R.: Adv. in Alicyclic Chem. *1,* 284 (1966); Beckwith, A. L. J., MTP International Review of Science, Vol. *10,* 1 (1973); Wood, D. E. et al.: J. Am. Chem. Soc. *94,* 6241 (1972); Symons, M. C. R.: Tetrahedron Lett. *1973,* 207; Lisle, J. B., Williams, L. F., Wood, D. E.: J. Am. Chem. Soc. *98,* 227 (1976); Krusic, P. J., Meakin, P.: J. Am. Chem. Soc. *98,* 228 (1976); Krusic, P. J., Bingham, R. C.: J. Am. Chem. Soc. *98,* 230 (1976); Bonazzola, L., Leray, N., Roncin, J.: J. Am. Chem. Soc. *99,* 8348 (1977); McBride, J. M.: J. Am. Chem. Soc. *99,* 6760 (1977); Claxton, T. A., Platt, E., Symons, M. C. R.: Molecular Phys. *32,* 1321 (1976); Dyke, J. et al.: Phys. Sci. *16,* 197 (1977); C.A. *89,* 107103 (1978); Griller, D. et al.: J. Am. Chem. Soc. *100,* 6750, (1978). Giese, B., Beckhaus, H. D., Angew. Chem. *90,* 635 (1978); Angew. Chem. Int. Ed. Engl. *17,* 594 (1978). In any case, all results point to much weaker force constants of out of plane deformations for free radicals than for carbenium ions. Bulky substituents seem to increase the tendency for a planar geometry of a radical center as e.g., in 2.2.-di-t-butyl cyclopropyl radicals; cf. Malatesta, V., Forrest, D., Ingold, K. U.: J. Am. Chem. Soc. *100,* 7073 (1978)
15. c.f. Eliel, E. L., Stereochemistry of Carbon Compounds, 1st edition, p. 267 ff. New York: McGraw-Hill Book Co. 1962, Eliel, E. L. in Newman, M. S., Steric Effects in Organic Chemistry, p. 212ff. New York: Wiley 1965
16. The geometry at the transition states of the ionisation is close to the sp^2-state of the carbenium ion: cf. Arnett, E. M., Petro, C.: J. Am. Chem. Soc. *100,* 2563 (1978)
17. cf. Hammond, G. S.: J. Am. Chem. Soc. *77,* 334 (1955)
18. Rüchardt, C. et al.: Structure Reactivity-Relationships in the Chemistry of Aliphatic Free Radicals. XXIII. Internat. Congr. Pure and Appl. Chemistry, Vol. 4, p. 223. Special Lectures, London: Butterworths 1971
19. Overberger, C. G. et al.: J. Am. Chem. Soc. *75,* 2078 (1953)
20. Bonnekessel, J., Rüchardt, C.: Chem. Ber. *106,* 2890 (1973)
21. Hinz, J., Rüchardt, C.: Liebigs Ann. Chem. *765,* 94 (1972)
22. Applequist, D. E., Klug, J. H.: J. Org. Chem. *43,* 1729 (1978)

23. Schuh, H., et al.: Helv. Chim. Acta *57*, 2011 (1974)
24. Beckhaus, H. D., Schoch, J., Rüchardt, C.: Chem. Ber. *109*, 1369 (1976)
25a. Lorenz, P., Rüchardt, C., Schacht, E.: Chem. Ber. *109*, 1369 (1976); b. The thermal decomposition of cycloalkanepercarboxylates and their α-methyl- and α-phenyl derivatives was recently reinvestigated very carefully by Wolf, R. A., Migliore, M. J., Fuery, P. H., Gagnier, P. R., Sabeta, J. C., Trocino, R. J.: J. Am. Chem. Soc. *100*, 7867 (1978); Their results are in perfect agreement with the interpretation given earlier[25a)] although a slightly modified interpretation is offered, see also Nelsen, S. F., Peacock, V. E., and Kesse, C. R. J. Am. Chem. Soc. *100*, 7017 (1978)
26. Schulz, A., Nguyen-Tran-Giac, Rüchardt, C.: Tetrahedron Lett. *1977*, 845
27. Tidwell, T. T.: Tetrahedron *34*, 1855 (1978), see also Ziebarth, M. and Neumann, W. P. Liebigs Ann. Chem. *1978*, 1765
28a. Bandlish, B. K. et al.: J. Am. Chem. Soc. *97*, 5856 (1975); b. Garner, A. W. et al.: J. Am. Chem. Soc. *97*, 7377 (1975); c. Prochazka, M.: Collect. Czechoslov. Chem. Commun. *41*, 1557, (1976); d. Duismann, W. et al.: Liebigs Ann. Chem. *1976*, 1820; Nguyen-Tran-Giac, Rüchardt, C.: Chem. Ber. *110*, 1095 (1977)
29a. Overberger, C. G., et al.; J. Am. Chem. Soc. *76*, 6185 (1954); b. Overberger, C. G., DiGiulio, A. V. J. Am. Chem. Soc. *81*, 2154 (1959); c. Lim, D.: Collect. Czechoslov. Chem. Commun. *33*, 1122 (1968)
30a. Overberger, C. G., DiGiulio, A. V.: J. Am. Chem. Soc. *81*, 1194 (1959); b. Prochazka, M., Rejmanova, P., Ryba, O.: Collect, Czechoslov. Chem. Commun. *39*, 2404 (1974); c. Prochazka, M., Ryba, O., Lim, D.: Collect. Czechoslov. Chem. Commun. *36*, 2640, 3650 (1971); d. Kovacic, P. et al.: J. Org. Chem. *34*, 3312 (1969); e. Gohen, S. G., Groszos, S. J., Sparrow, D. B.: J. Am. Chem. Soc. *72*, 3947 (1950); f. Brooks, B. W., Dainton, F. S., Ivin, K. I.: Trans. Farad. Soc. *61*, 1437 (1965)
31. Newman, M. S.: in Steric Effects in Organic Chemistry 1st edition, p. 206, New York: Wiley 1956
32. Prochazka, M.: Collect. Czechoslov. Chem. Commun. *42*, 2394 (1977)
33. Duismann, W., Rüchardt, C.: Chem. Ber. *106*, 1083 (1973)
34. Schreiner, K., Berndt, A.: Angew. Chem. *87*, 285 (1975) Angew. Chem. Int. Ed. Engl. *14*, 366 (1975)
35a. Koenig, T.: in Free Radicals, Kochi, J. K. (ed.) 1st edition, Vol. 1, p. 113, New York: Wiley Interscience 1973; b. Hinz, J., Oberlinner, A., Rüchardt, C.: Tetrahedron Lett. *1973*, 1975; c. Engel, P. S., Bishop, D. J.: J. Am. Chem. Soc. *97*, 6754 (1975); d. Koga, G., Anselme, J. P. in The Chemistry of the Hydrazo, Azo and Azoxy Groups, Patai, S. (ed.), Vol. 2, 1st edition, p. 861, New York: Interscience 1975
36. cf. Green, J. G., Porter, N. A.: J. Am. Chem. Soc. *99*, 1264 (1977), Suehiro, T., et al.: Bull. Chem. Soc. Jap. *50*, 3325 (1977). Porter, N. A., Dubay, G. R., Green, J. G.: J. Am. Chem. Soc. *100*, 920 (1978); Pryor, W. A., Smith, K.: J. Am. Chem. Soc. *89*, 1741 (1967)
37. Schulz, A., Rüchardt, C.: Tetrahedron Lett. *1977*, 849
38a. Ernst, J. A., Thankachan, C., Tidwell, T. T.: J. Org. Chem. *39*, 3614 (1974); b. Duismann, W., Rüchardt, C.: Liebigs Ann. Chem. *1976*, 1834
39a. Engler, E. M., Andose, J. D., Schleyer, P. v. R.: J. Am. Chem. Soc. *95*, 8005 (1973); b. Schleyer, P. v. R., Williams, J. B., Blanchard, K. R.: J. Am. Chem. Soc. *92*, 2377 (1970)
40. Andose, J., Mislow, K.: J. Am. Chem. Soc. *96*, 2168 (1974)
41. Hellmann, G., Beckhaus, H. D., Rüchardt, C.: Chem. Ber. *1979* in print
42. Tsang, W.: J. Chem. Phys. *44*, 4283 (1966)
43. Beckhaus, H. D., Rüchardt, C.: Chem. Ber. *110*, 878 (1977)
44. Rüchardt, C., Winiker, R.: unpublished
45. Rüchardt, C. et al.: Angew. Chem. *89*, 913 (1977); Angew. Chem. Int. Ed. Engl. *16*, 875 (1977)
46. Tsang, W.: J. Chem. Phys. *43*, 352 (1965)
47. Baxter, S. G. et al.: J. Am. Chem. Soc. submitted
48. Beckhaus, H. D., Hellmann, G., Rüchardt, C.: Chem. Ber. *111*, 3764 (1978)
49. Beckhaus, H. D., Hellmann, G., Rüchardt, C.: Chem. Ber. *111*, 72 (1978)

50. Berces, T., Seres, L., Manta, F.: Acta Chim. Acad. Sci. Hung. *71*, 31 (1973)
51. Burcat, A. et al.: Int. J. Chem. Kinetics *5*, 345 (1973)
52. Dissertation Hellmann, G., Univ. Freiburg 1977
53. Diplomarbeit Kratt, G., Univ. Freiburg 1976
54. Stein, S. E., Golden, D. M.: J. Org. Chem. *42*, 839 (1977)
55. Dissertation Dempewolf, E., Univ. Freiburg 1977
56. Bartlett, P. D., McBride, M. J.: J. Am. Chem. Soc. *87*, 1727 (1965)
57. Sargent, G. D.: in G. Olah, Schleyer, P. v. R., Carbonium Ions III, New York: Wiley-Interscience *1972*, 1099
58. Golzke, V. et al.: Nouv. J. Chim. *2*, 169 (1978)
59. Dissertation Golzke, V., Univ. Freiburg 1977; s. a. Heine, H. G. et al.: J. Org. Chem. *41*, 1907 (1976) for the corresponding photochemical formation of bridgehead radicals
60. Bunce, N. J., Hadley, M.: J. Org. Chem. *39*, 2271 (1974)
61. McKean, D. C.: Chem. Soc. Rev. *7*, 399 (1978)
62. Bartlett, P. D. et al.: Acc. Chem. Res. *3*, 177 (1970)
63. Gruselle, M., Lefort, D.: Tetrahedron *32*, 2719 (1976)
64. Fort, R. C., Schleyer, P. v. R.: Adv. in Alicyclic Chem. *1*, 284 (1966); Koch, V. R., Gleicher, G. J.: J. Am. Chem. Soc. *93*, 1657 (1971)
65. Tabushi, I., Kojo, S., Fukunishi, K.: J. Org. Chem. *43*, 2370 (1978)
66. Mazur, Y., Cohen, Z.: Angew. Chem. *90*, 289 (1978); Angew. Chem. Int. Ed. Engl. *17*, 281 (1978)
67. Opeida, I. A., Timokhin, V. I.: Ukr. Khim. Zh. *44*, 187 (1978); C. A. *88*, 189609 (1978); Koshel, G. N. et al.: Zh. Org. Khim. (USSR) *14*, 534 (1978)
68. Cristol, S. J. et al.: J. Org. Chem. *41*, 1919 (1976)
69. Russell, G., Free Radicals, Kochi, J. K. (ed.) Vol. 1, p. 312 1st edition, New York: Wiley 1973; s. a. Breslow, R. et al.: J. Am. Chem. Soc. *94*, 3276 (1972)
70a. Deno, N. C.: in Methods in Free Radical Chemistry (E. S. Huyser) Vol. 3, New York: Dekker 1972; Deno, N. C., Pohl, D. G.: J. Org. Chem. *40*, 380 (1975); J. Am. Chem. Soc. *96*, 6680 (1974); b. Minisci, F.: Synthesis *1973*, 1; c. Bernardi, R., Galli, R., Minisci, F.: J. Chem. Soc. B, *1968*, 324; d. Minisci, F.: Topics Curr. Chem. *62*, 1 (1976); e. Johnson, R. A., Green, F. D.: J. Org. Chem. *40*, 2192 (1975)
71. Deno, N., Fishbein, R., Wyckoff, J. C.: J. Am. Chem. Soc. *93*, 2065 (1971)
72. Breslow, R. et al.: J. Am. Chem. Soc. *94*, 3277 (1972)
73. Breslow, R.: Chem. Soc. Rev. *1*, 553 (1972); Breslow, R. et al.: J. Am. Chem. Soc. *100*, 1213 (1978)
74. Breslow, R. et al.: J. Am. Chem. Soc. *99*, 905 (1977)
75. Rotman, A., Mazur, Y.: J. Am. Chem. Soc. *94*, 6228 (1972); Mazur, Y.: Pure Appl. Chem. *41*, 145 (1975)
76. Barton, D. H. R. et al.: J. Am. Chem. Soc. *98*, 3036 (1976)
77. Akhtar, M.: Adv. Photochemistry *2*, 263 (1964); Mihailovic, M. L., Cekovic, Z.: Synthesis *1970*, 209; Kalvoda, J., Heusler, K.: Synthesis *1971*, 501; Heusler, K.: Heterocycles *3*, 1035 (1975); Hesse, R. H., Adv. Free Rad. Chem. *3*, 83 (1969)
78. Wilt, J. W.: in Kochi, J. K., Free Radicals, Vol. 1, p. 333, New York: Wiley 1973
79. Herwig, K., Lorenz, P., Rüchardt, C.: Chem. Ber. *108*, 1421 (1975)
80. Giese, B.: Angew. Chem. *88*, 159, 161, 723 (1976); Angew. Chem. Int. Ed. Engl. *15*, 173, 174, 688 (1976)
81. Private communication of Prof. Giese, Darmstadt
82. Fujita, T., Takayama, C., Nakajima, M.: J. Org. Chem. *38*, 1623 (1973)
83. Giese, B.: Angew. Chem. *89*, 162 (1977); Angew. Chem. Chem. Int. Ed. Engl. *16*, 125 (1977)
84. Beckhaus, H. D.: Angew. Chem. *90*, 633 (1978); Angew. Chem. Int. Ed. Engl. *17*, 592 (1978)
85. Giese, B., Beckhaus, H. D.: Angew. Chem. *90*, 635 (1978) Angew. Chem. Int. Ed. Engl. *17*, 594 (1978)
86. Giese, B., Stellmach, J.: Tetrahedron Lett. *1979*, 857
87. Szeimies, G. et al.: Chem. Ber. *111*, 1922 (1978); Dietz, P., Szeimies, G., Chem. Ber. *111*, 1938 (1978)

88. Davies, A. G. et al.: J. Organometal. Chem. *118*, 289 (1976)
89. For a summary see Davies, D. I. in MTP International Review of Science, Vol. *10*, p. 49, London, Butterworths 1973; Abell, P. I., Free Radicals, Kochi, J. K. (ed.) Vol. *2*, p. 63, New York: Wiley 1973
90. Walling, C.: Free Radicals in Solution, New York: Wiley 1957. For a recent MINDO/3 study supporting this interpretation see Dewar, M. J. S. and Olivella, S., J. Am. Chem. Soc. *100*, 5290 (1978)
91. Summaries at Beckwith, A. L. J., Essays in Free Radical Chemistry Norman, R. O. C. ed. Chemical Society, Special Publ. *1970*, 239; Julia, M.: Pure and Appl. Chem. *15*, 167 (1967); Nonhebel, D. C., Walton, J. C.: Free Radical Chemistry, p. 533ff. Cambridge: University Press 1974
92. Clark, D. T., Scanlan, J. W., Walton, J. C.: Chem. Phys. Lett. *55*, 102 (1978)
93. Tedder, J. M., Walton, J. C.: Acc. Chem. Research *9*, 183 (1976)
94. Brown, H. C.: Organic Synthesis via Boranes, New York: Wiley 1975; Davies, D. I., Parrott, J. M., Free Radicals in Organic Synthesis, Berlin-Heidelberg-New York: Springer 1978
95. Giese, B., Meister, J.: Chem. Ber. *110*, 2588 (1977)
96. Jenkins, A. D.: Adv. in Free Radical Chem. *2*, 139 (1967)
97. Bonacic-Koutecky, V., Koutecky, J., Salem, L.: J. Am. Chem. Soc. *99*, 842 (1977)
98. Riemenschneider, K. et al.: Tetrahedron Lett. *1979*, 185; Riemenschneider, K. et al.: Tetrahedron Lett. *1979*, 189
99. Giese, B., Meixner, J.: Tetrahedron Lett. *1977*, 2779 and Ref.[8)]
100. Capka, M., Chvalovsky, V.: Collect. Czechoslov. Chem. Commun. *33*, 2872 (1968)
101. Dissertation Müller, H. J.: Univ. Freiburg 1977
102. Low, H. C., Tedder, J. M., Walton, J. C.: Int. J. Chem. Kinet. *10*, 325 (1978)
103. Giese, B., Zwick, W.: Angew. Chem. *90*, 62 (1978); Angew. Chem. Int. Ed. Engl. *17*, 66 (1978)
104. Szwarc, M., Binks, J. H.: Theoretical Organic Chemistry Kekule Symposium 1958, p. 262, London: Butterworths 1958
105. Stefani, A. P., Chuang, L. Y. Y., Todd, H. E.: J. Am. Chem. Soc. *92*, 4168 (1970)
106. Yip, R. W. et al.: J. Phys. Chem. *82*, 1194 (1978)
107. Cessna, A. J. et al.: J. Am. Chem. Soc. *99*, 4044 (1977)
108. Giese, B., Meister, J.: Angew. Chem. *89*, 178 (1977); Angew. Chem. Int. Ed. Engl. *16*, 178 (1977)
109. Caronna, T. et al.: Tetrahedron *33*, 793 (1977); Citterio, A. et al.: J. Am. Chem. Soc. *99*, 7960 (1977)
110. Giese, B., Jay, K.: Chem. Ber. *110*, 1364 (1977)
111. Perkins, M. J.: Free Radicals, Kochi, J. K. (ed.), Vol. 2, p. 231, 1st edition, New York: Wiley 1973
112. Minisci, F.: Topics Curr. Chem. Vol. 62, Heidelberg: Springer 1970, p. 3; Sosnovsky, G., Rawlinson, D. J., Adv. Free Radical Chem. *4*, 203 (1972)
113. Chow, Y. L. et al.: Chem. Rev. *78*, 243 (1978)
114. Kornblum, N.: Angew. Chem. *87*, 797 (1975), Angew. Chem. Int. Ed. Engl. *14*, 734 (1975); Kornblum, N.: Pure and Appl. Chem. *4*, 81 (1971); Kornblum, N.: J. Am. Chem. Soc. *100*, 289 (1978)
115. Bunnett, J. F.: J. Chem. Educ. *51*, 313 (1974); Bunnett, J. F.: Acc. Chem. Research *11*, 431 (1978)
116. Norris, R. K., Randler, D.: Austr. J. Chem. *29*, 2621 (1976)
117. Gibian, M. J., Corley, R. C.: Chem. Rev. *73*, 441 (1973); Stein, S. E., Rabinovitch, B. S.: Int. J. Chem. Kinetics *7*, 531 (1975)
118. Schuh, H. H., Fischer H.: Helv. Chim. Acta *61*, 2130, 2463 (1978)
119. Pershin, A. D. et al.: Collect. Czechoslov. Chem. Commun. *43*, 1349 (1978)
120. Eichin, K. H. et al.: Angew. Chem. *90*, 987 (1978); Angew. Chem. Int. Ed. Engl. *17*, 934 (1978)

Received March 22, 1979

Silylated Synthons

Facile Organic Reagents of Great Applicability

L. Birkofer and O. Stuhl

Institut für Organische Chemie der Universität Düsseldorf, Universitätsstraße 1, D-4000 Düsseldorf

Contents

Abbreviations

Cp: cyclopentadienyl-,
DCB: dicyclohexyl borane
DIBAL: di-isobutylaluminium hydride
DIBATO: bis(dibutylacetoxytin)oxide
DME: dimethoxyethane
DMSO: dimethyl sulfoxide
Et: ethyl-,
HMCTS: hexamethylcyclotrisilazane
HMPT: hexamethylphosphoric triamide
LDA: lithiumdiisopropylamide
MCPBA: meta-chloro-perbenzoic acid
Me: methyl-,
MES: mesityl-,
TES: triethylsilyl-,
THF: tetrahydrofurane
TMS: trimethylsilyl-,
TPS: triphenylsilyl-,
Z: carbobenzoxy (cbo)

A Introduction

In the last two decades the importance of organosilicon chemistry has greatly increased. Especially, silylated synthons – synthesized by a great variety of silylation reactions – have won wide appreciation and became a highly valued and often used tool in every preparative chemist's hand.

The spectrum of possible applications has expanded[1–16] and the number of publications in this field has steadily increased year by year. In this review we shall describe some of the main contemporary fields of use for silylated synthons and silylation techniques.

B Reactions of Organosilanes with Unsaturated Hydrocarbons

Organosilanes react with unsaturated hydrocarbons via two different types of reaction:

a) Addition and elimination reactions;
b) electrophilic substitution under Friedel-Crafts conditions where the silyl moiety primarily has protecting and/or activating function.

1 Hydrosilylation

From the turn of the century on, the principal route for linking a carbon-silicon bond in synthesizing variously substituted silanes was the classical "Grignard method", first introduced by Kipping[17, 18] and Dilthey[19]. Unfortunately, this reaction was mainly limited to obtain (organo-) saturated organosilanes.

During the last twenty years a very interesting addition of silanes across multiple bonds became familiar in organosilicon chemistry – hydrosilylation.

Early reports stated that the course of reaction is strongly dependent on the reaction conditions (i. e. the employed catalyst)[20–28]. Benkeser[20–22] and his co-workers intensively investigated hydrosilylation of monosubstituted acetylenes *1* [R = i-prop- (*1a*) and t-But- (*1b*)].

He showed that t-butyl-acetylene (*1b*) adds trichlorosilane (*2*) via cis addition or trans addition to yield the trans products (*3a, b*) and cis products (*4a, b*), respectively, depending on the kind of catalyst (Scheme 1).

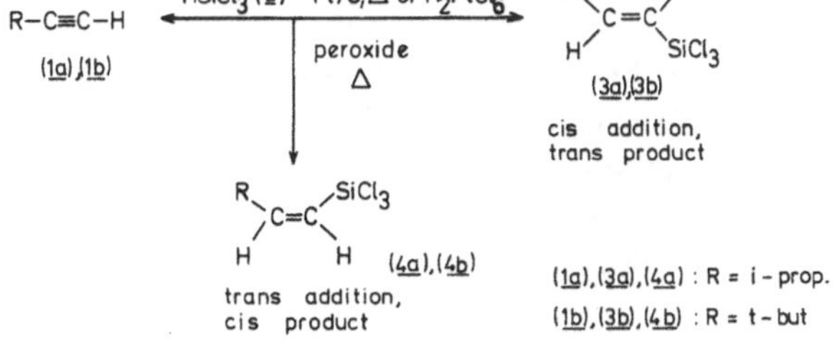

R–C≡C–H (1a),(1b) —[$HSiCl_3$ (2), Pt/C,Δ or H_2PtCl_6]→ (R)(H)C=C(H)($SiCl_3$) (3a),(3b) cis addition, trans product

R–C≡C–H (1a),(1b) —[$HSiCl_3$ (2), peroxide, Δ]→ (R)(H)C=C(H)($SiCl_3$) (4a),(4b) trans addition, cis product

(1a),(3a),(4a) : R = i-prop.
(1b),(3b),(4b) : R = t-but

Scheme 1

If platinum/carbon or hexachloroplatinic acid is taken as catalyst the trans products *3a, b* are obtained stereospecifically via cis addition whereas peroxide catalysis leads to the cis-products *4a, b* under trans addition with equal selectivity.

Later, Tamao, Kumada and co-workers[23] have shown that a bis-silylated ethene can be achieved if the hydrosilylation takes place in presence of Ni-II complexes: 3-hexyne (*5*) gives 3,4-bis(trichlorosilyl)-3-hexene (*6*) generating hydrogen (Scheme 2).

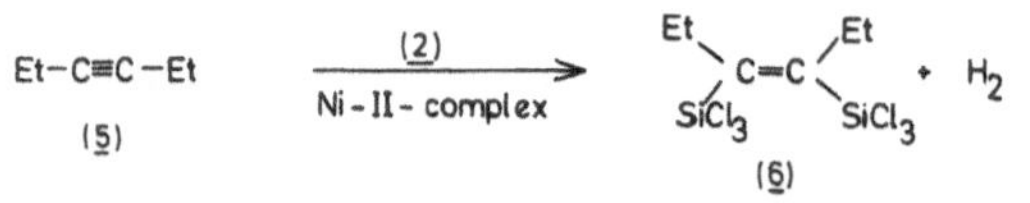

Et–C≡C–Et (5) —(2), Ni-II-complex→ (Et)(SiCl₃)C=C(Et)(SiCl₃) (6) + H_2

Scheme 2

Generally, it can be said that predominantly

a) the hydrosilylation of simple olefins and acetylenes places the silicon atom at the less substituted carbon atom

b) via catalysts and reaction conditions a stereospecific course of reaction can be obtained. Furthermore, as a very positive side effect, asymmetrical hydrosilylation can be accomplished if chiral catalysts are employed[24–27].

Quite remarkable are the successful attempts to synthesize C-silyl sugars[28] directly at the glucose ring by means of hydrosilylation of unsaturated sugars – in contrast to previous negative forecasts[29] ("nicht direkt am Glucosering"[29]).

To concluding this section a novel route[30] for synthesizing symmetrical bis(trimethylsilyl)ethene has to be explained. Starting with monotrimethylsilylacetylene (*7*), a hydrosilylation of *7* by means of chlorodimethylsilane (*8*) and hexachloroplatinic acid yields the trans-1,2-bis(silyl)ethene *9*, which is converted into the trans-1,2-bis(trimethylsilyl)ethene (*10*) by subsequent alkylation with the Grignard reagent CH_3MgJ (Scheme 3).

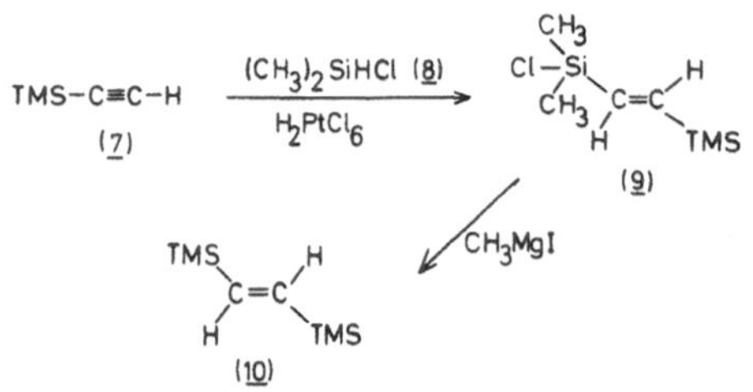

TMS–C≡C–H (7) —$(CH_3)_2SiHCl$ (8), H_2PtCl_6→ Cl–Si(CH_3)$_2$–CH=CH–TMS (9) —CH_3MgI→ TMS–CH=CH–TMS (10)

Scheme 3

2 Acetylenic Silanes

The stability of the -C≡C-Si⩽ bond has been known for a long time[31, 32]. But on the other hand they are reactive compounds which undergo either – as precursors to vinylsilanes – various types of addition reactions or – as only silyl-protected acetylenes – an electrophilic substitution under Friedel-Crafts conditions in presence of catalytic amounts of Lewis acids[33]. The $-SiR_3$ moiety has a highly useful protecting and/or activating function.

One of the classical additions is the reaction of phenyl-TMS-acetylene (*11*) or bis(trimethylsilyl)acetylene (*12*) with trimethylsilylazide (*14*) to form 1,4-bis-(trimethylsilyl)-5-phenyltriazole (*15*) and 1,4,5-tris(trimethylsilyl)-triazole (*16*), respectively[33]. The photochemical addition[34] of *7* and *12* with maleic anhydride give mono- and bis(trimethylsilyl)cyclobutene-3,4-dicarboxylic anhydrides (*17*) and (*18*), whereas *12* with oxalyl chloride in presence of catalytic amounts of aluminium chloride leads to the heterocyclic 2,2-dichloro-4,5-bis(trimethylsilyl)-2,3-dihydro-furan-3-one (*19*)[35]. If *11* is treated with dicyclohexylborane (DCB), the corresponding 1-phenyl-2-trimethylsilyl-2-boranyl-ethene (*20*) is formed which can be converted by subsequent oxidation with alkaline/hydrogen peroxide to phenylacetic acid (*21*)[36]. Quite remarkable is the course of reaction when an alkyl-trimethylsilyl-acetylene (e. g. *13*) is treated with di-isobutylaluminumhydride (Dibal) as the reaction medium influences the stereochemistry: in n-hexane/methylpyrrolidine the trans product (*22a*), whereas in n-hexane the cis product (*22b*) is obtained. In the following step, alkylation by means of methyllithium and alkylhalide is possible, leading to the corresponding ethenes (*23a*) and (*23b*)[37]. The same orientation occurs when the homologous triethylsilylacetylene (*13a*) is taken[38]. An interesting parallel is shown[39] when sodiumtriethylborohydride is employed, the sodium salt of trimethylsilyl-triethyl-boranyl-acetylene (*24*) is achieved; further reaction with electrophilic reagents (e. g. chlorodimethylether, chlorotrimethylsilane, chlorodiphenylphosphine) leads to the variously substituted ethenes (*25a–c*). Recent papers describe a) the combination of hydroboration with subsequent treatment with base/copper halide and then with alkylhalide[40] or b) the direct employment of organocuprates/magnesium halide reagent intermediates[41, 42] and furthermore with an organohalide to obtain manifold substituted ethenes (*26*) and (*27*). 2,3,6,7-tetrakis(trimethylsilyl)-naphthalene (*28*) has been synthesized in a Co-carbonyl complex catalyzed reaction[43] of two equivalents of *12* with one equivalent of 3-trimethylsiloxy-1,5-hexa-di-yne (Scheme 4).

Another reaction pathway of great synthetic usefulness is the addition of halogen across bis(trimethylsilyl)ethyne (*12*)[30] to isolate trans-1,2-dihalogeno-1,2-bis(trimethylsilyl)ethenes *29* and *30*. Further halogenation leads to symmetrical (*31, 32*) and nonsymmetrical (*33*) ethane compounds; the pyrolysis of *31* yields 1,1,2-trichloro-2-(trimethylsilyl)ethene (*34*) (Scheme 5).

Quite important are silylated acetylenes[33] for the synthesis of poly-ynes and a large number of variously substituted acetylenes[33, 44–52].

Sometimes the triethylsilyl grouping is more suitable for poly-acetylene synthesis because of its greater stability, namely in the Cadiot-Chodkiewicz coupling reaction[44] of a halogenoacetylene with unprotected 1-phenyl-buta-1,3-diyne (*39*) [e. g. (*39*) → (*42*)] or in the Hay-coupling reaction[45] of semi-silylated acetylenes by means of oxygen [e. g. (*43*) → (*46*)].

Later Walton and his co-workers[46] could show that 2 equivalents of 2-trimethyl-silyl-ethyne-magnesiumbromide couple with cyclooctatetraene dibromide to form 1,12-bis(trimethylsilyl)dodeca-3,5,7,9-tetraene-1,11-diyne (*47*).

Under Friedel-Crafts-type conditions, many electrophilic displacements of the silyl moiety occur, so 1-phenylsulfonyl-2-TMS-ethyne (*49*) can be prepared from *12* with benzenesulfonyl chloride in presence of catalytic amounts of aluminum chlo-

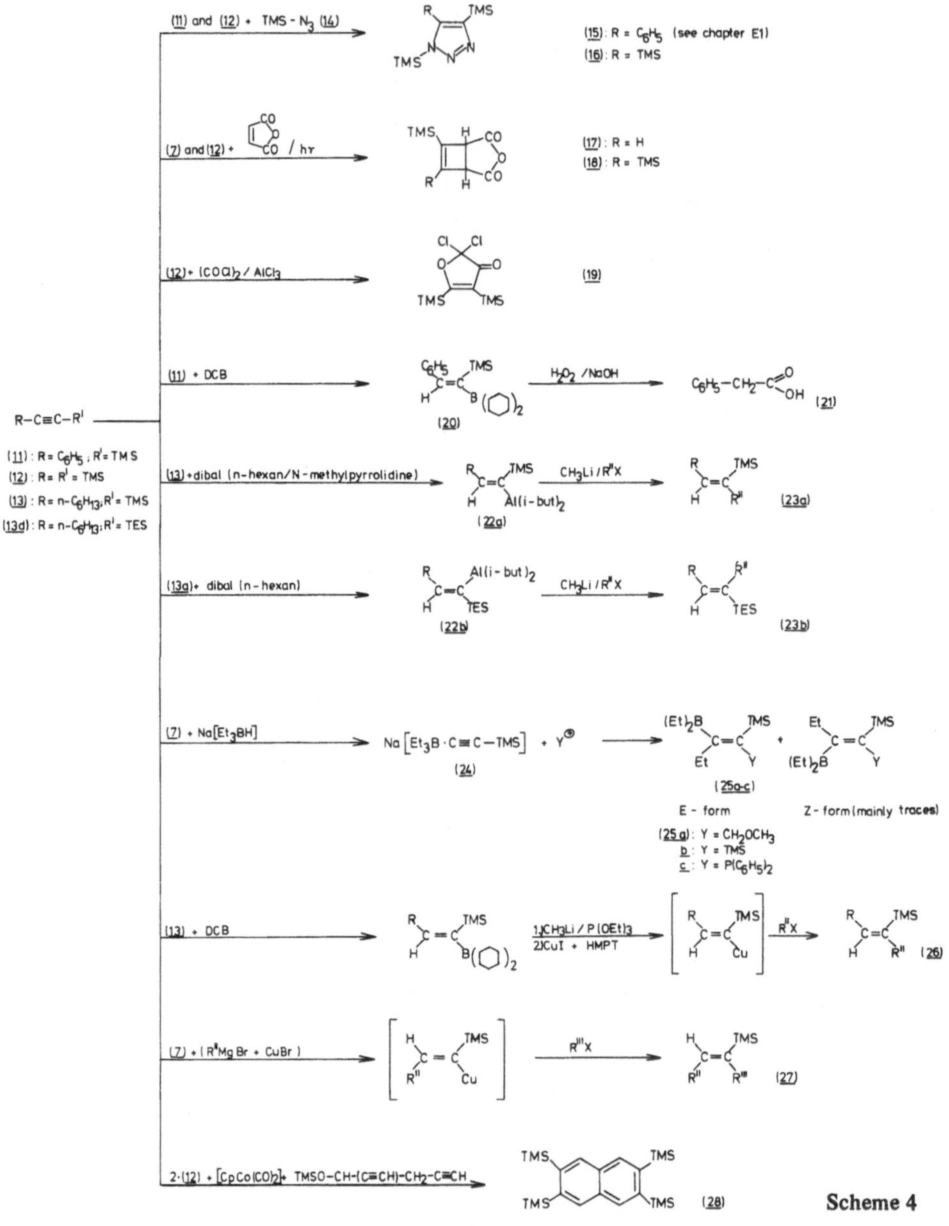

Scheme 4

ride[47]. With an interhalogen compound, e. g. iodine chloride, *12* can be transformed to 1-iodo-2-TMS-ethyne (*50*).

Similarly, carbonyl derivatives such as acid chlorides[33, 51], acid anhydrides[35] or N-substituted carbamoylchlorides[52] are highly reactive reagents for performing an electrophilic attack. With phthalic anhydride and 1,4-bis(TMS)-buta-1,3-diyne the corresponding o-(5-TMS-penta-2,4-diyne-1-one-1-yl)benzoic acid (*51*) is formed[35].

TMS−C≡C−TMS (12) —X_2/$ZnCl_2$→ TMS(X)C=C(X)TMS

	X
(29)	Cl
(30)	Br

↓ Y_2/$ZnCl_2$

$Cl_2C{=}C(Cl)TMS$ (34) ←Δ, −TMS−Cl— (31); TMS−CXY−CXY−TMS

	X	Y
(31)	Cl	Cl
(32)	Br	Br
(33)	Cl	Br

Scheme 5

Furthermore unsaturated aldehydes can be obtained by the reaction of *12* with e. g. cyclohexane carboxylic acid chloride, subsequent alkaline cleavage of the remaining silyl grouping and reduction with sodium borohydride give 3-hydroxy-3-cyclohexyl-1,1-dimethoxypropane (*54*) which yields 3-cyclohexyl-acrolein (*55*)[51] via acid treatment under dehydratation.

Equally, in presence of $AlCl_3$, if N,N-diethyl-carbamoyl chloride is applied to an 1-alkyl-2-TMS-acetylene, the corresponding propargylic acid amide *56* can be isolated[52] (see Scheme 6).

Very closely related to the above described synthesis[52] are the successful attempts of Metcalf and co-workers to prepare α-acetylenic α-amino acids[53] following up from previous results[54, 55].

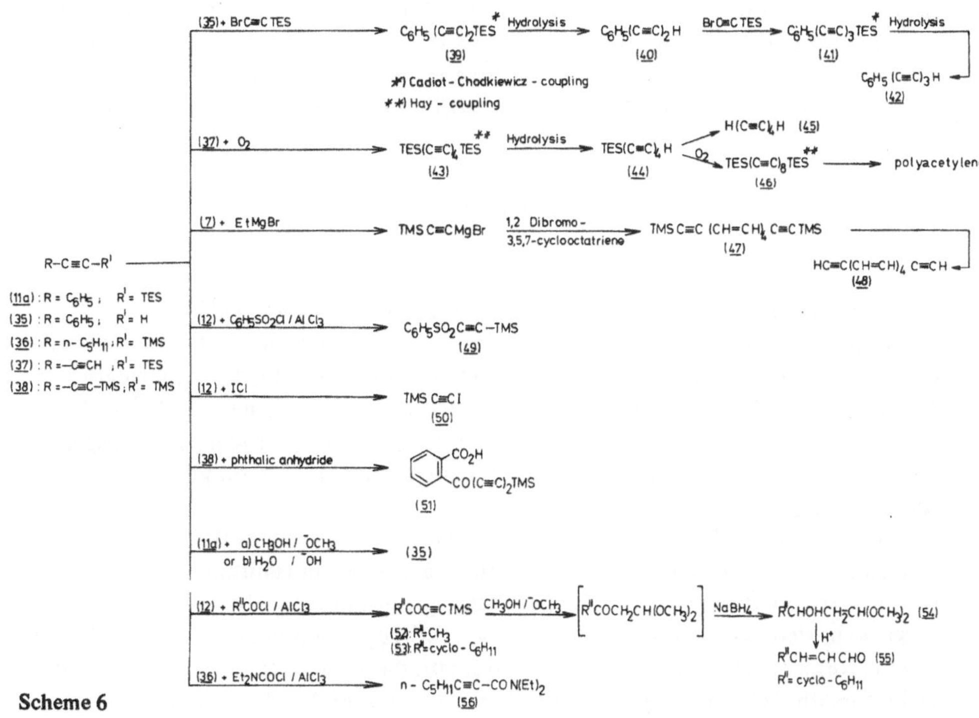

Scheme 6

Via an amidoalkylation of bis(trimethylsilyl)ethyne (*12*) with methyl 2-chloro-N-ethoxycarbonylglycinate (*57*) under the influence of aluminum chloride the corresponding N-(ethoxycarbonyl)-α,α-TMS-ethynyl-glycinate (*58*) is isolated which is converted by means of lithium-di-isopropylamide (LDA) into a carbanion that reacts with alkylhalide to the substituted glycinate *59*. Finally, after alkaline hydrolysis the unprotected α-acetylenic-α-aminoacid (*60*) (e. g. R = Benzyl: α-ethynyl-α-phenylalanine is then obtained (Scheme 7).

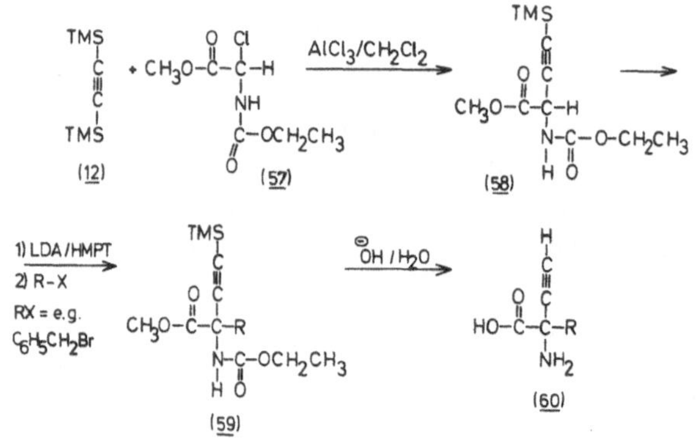

TMS–C≡C–TMS (12) + CH_3O–C(=O)–CH(Cl)–NH–C(=O)–OCH_2CH_3 (57) —$AlCl_3/CH_2Cl_2$→ CH_3O–C(=O)–CH(C≡C–TMS)–N(H)–C(=O)–O–CH_2CH_3 (58) → 1) LDA/HMPT 2) R–X (RX = e.g. $C_6H_5CH_2Br$) → CH_3O–C(=O)–C(R)(C≡C–TMS)–N(H)–C(=O)–OCH_2CH_3 (59) —$^{\ominus}OH/H_2O$→ HO–C(=O)–C(R)(C≡C–H)–NH_2 (60)

Scheme 7

3 Vinylsilanes

Vinylsilanes are proper compounds which have won great appreciation as facile intermediates for many sorts of application. They are prepared either by addition reactions of acetylenes (vide supra) or by elimination reactions of saturated organosilanes. The main reactions, carried out with vinylsilanes, are addition reactions to obtain saturated organosilanes or electrophilic substitutions of the silyl group under Friedel-Crafts conditions where a rapid cleavage of the silyl moiety occurs (Scheme 8).

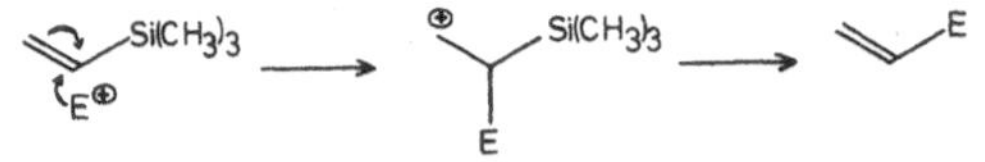

CH_2=CH–$Si(CH_3)_3$ + $E^{\oplus}$ → $^{\oplus}$CH(–CH_2E)–$Si(CH_3)_3$ → CH_2=CH–E

Scheme 8

This can be illustrated by the acylation of linear[56, 58] and cyclic[57] vinylsilanes. It is quite simple to prepare methylvinylketone (*69*)[56] or 1-methyl-6-TMS-hexa-2,5-diene-4-one (*70*)[58]. In the latter case the starting material is bis(trimethylsilyl)-ethene (*62*) which can either be achieved via hydrosilylation (vide supra)[30] or by the more classical approaches[59–61]. Furthermore, because of the high stereospecifity, compounds like E- cinnamaldehyde (*71*) are obtainable in good yields and with superb optical purity[62].

However, if a halogen is attached at the same carbon atom as the TMS-group is situated, the attack of a powerful electrophilic agent like butyllithium takes a different direction, it removes the halogen forming a lithiumderivative which then undergoes the Peterson-reaction (vide infra) if a carbonyl compound is added[64, 67]. Via the former method E-1-phenyl-2-chloromethyl-ethene (*72*) and E-1-cyclohexyl-2-chloromethyl-ethene (*73*) can be synthesized. E-1-TMS-1-octene can be converted stereospecifically into E-1-deutero-1-octene (*74*) under acid catalysis in presence of iodide[65]. Very important and of great synthetic potential are halogenation procedures[30, 66, 67, 68, 69, 71].

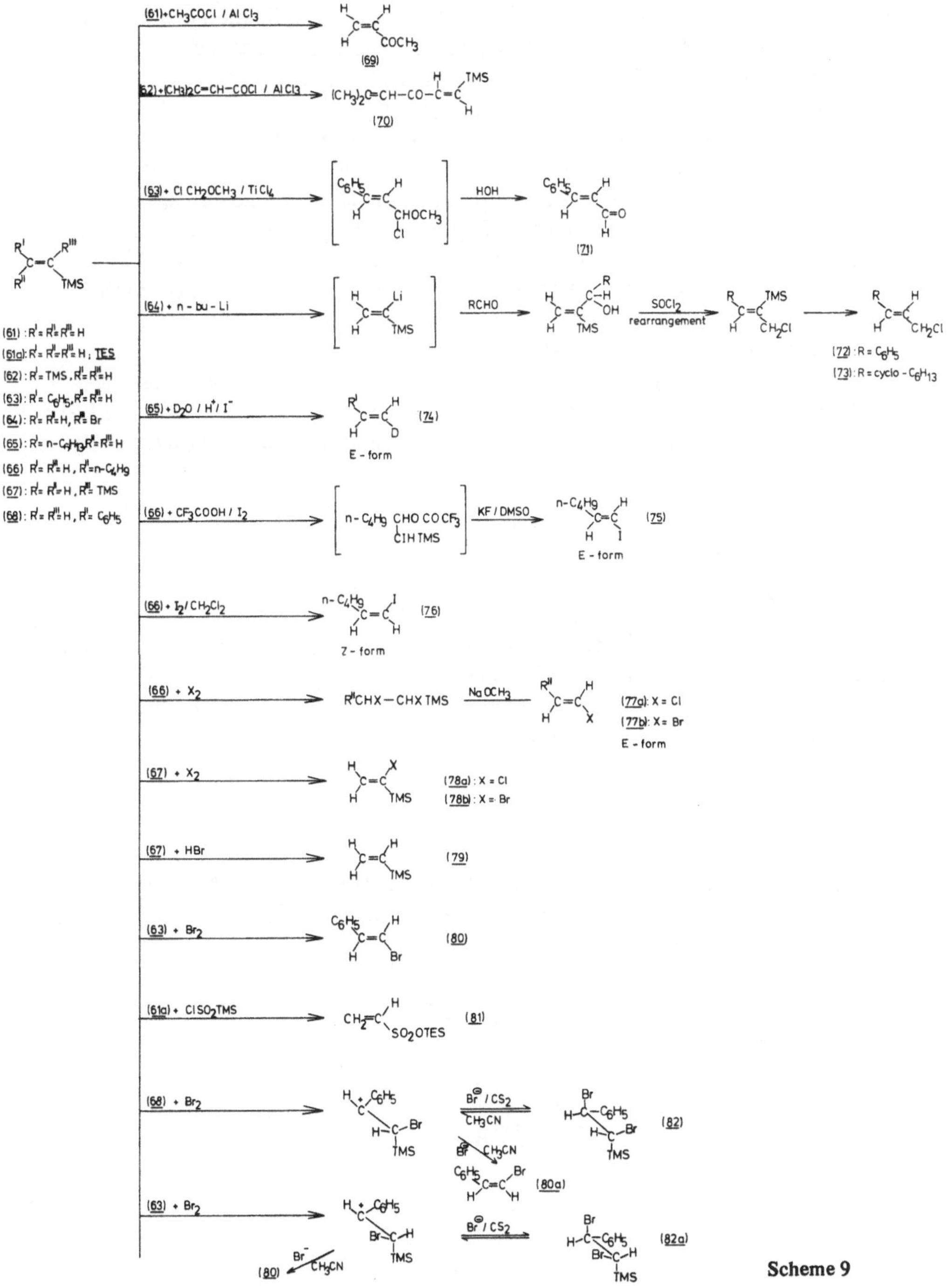

(61): R' = R'' = R''' = H
(61a): R' = R'' = R''' = H; TES
(62): R' = TMS, R'' = R''' = H
(63): R' = C_6H_5, R'' = R''' = H
(64): R' = R''' = H, R'' = Br
(65): R' = n-C_6H_{13}, R'' = R''' = H
(66) R' = R''' = H, R'' = n-C_4H_9
(67): R' = R''' = H, R'' = TMS
(68): R' = R''' = H, R'' = C_6H_5

(72): R = C_6H_5
(73): R = cyclo-C_6H_{13}

(77a): X = Cl
(77b): X = Br
E-form

(78a): X = Cl
(78b): X = Br

Scheme 9

Via addition and elimination in two subsequent steps a conversion of the stereochemistry of the vinylsilanes is obtained. A Z-substrate yields an E-product (e. g. *66* → *75*, *66* → *77*). This has great impact on chlorination and bromination reactions. The iodination however, in a highly polar medium, for instance trifluoroacetic acid, shows an equal course of reaction (i. e. Z → E) whereas the iodination in a nonpolar solvent (i. e. methylene chloride[66]) leaves the stereochemistry untouched.

Furthermore, the stereochemical course of halogenation procedures was intensively studied by Koenig and Weber[71] in case of cis (*68*)- and trans-2-TMS-styrene (*63*). Their results showed the strong stereochemical specificity in the reactions of vinylsilanes with bromine. Other reports[67, 68, 69] confirm the highly stereospecific course of reaction (e. g. *67* → *78*, *63* → *80*) during the electrophilic displacement. However, it must be noted that the use of hydrobromic acid with 1,1-bis(trimethylsilyl)ethene (*67*) leads to 1-TMS-ethene (*79*). Sulfonation is applicable, too[70]. If the sulfonation reagent Cl-SO_2-O-TMS is added to 1-triethylsilylethene (*61a*) the corresponding sulfonyl compound (*81*) can be prepared, but an exchange of TES against TMS is observed (Scheme 9).

Concluding this chapter a very interesting variant[72] of the Robinson-annelation[73] has to be described. In a Michael-type reaction under aprotic conditions Li-1-cyclohexene-1-olate (*83*) was employed to 2-TES-1-butene-3-one (*84*) (Scheme 10).

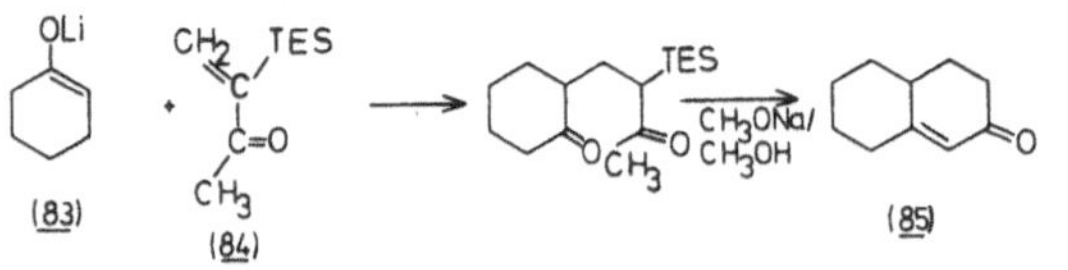

Scheme 10

At first, the vinylsilane was added but further treatment with sodium methoxide/methanol afforded desilylation and ring closure, so that $\Delta^{1,9}$-2-octalone (*85*) was isolated exclusively. This reaction demonstrates the activating properties of the trialkylsilyl function in a very impressive manner.

4 Epoxysilanes

The conversion of vinylsilanes into epoxysilanes was first published by Stork and Colvin[74] in 1971. They added meta-chloroperbenzoic acid (*87*) to 1-TMS-3-hydroxy-

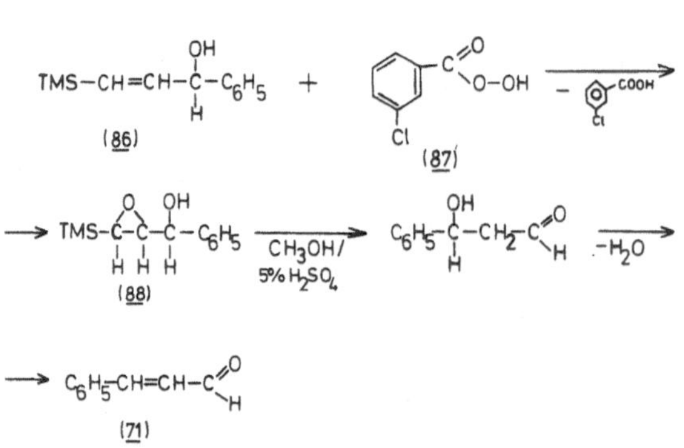

Scheme 11

3-phenyl-propene (*86*) and the corresponding epoxy compound (*88*) was formed. Subsequent acid catalyzed desilylation and ring opening plus elimination of water yields cinnamaldehyde (*71*) (Scheme 11). The α,β-epoxysilanes are very useful synthons of good reactivity to achieve a variety of products. The electrophilic-catalyzed ring opening of α,β-epoxy silanes is predominantly an α-cleavage, since the silyl moiety activates the α-position for a bimolecular nucleophilic attack of the reagent.

If trimethylsilylethene oxide (*89*) is treated with magnesium bromide[76], a brominating ring opening occurs to yield 2-bromo-2-TMS-ethanol (*100*). In a similar reaction with 1-propyl-2-TMS-oxirane (*91*) and 1 equivalent $MgBr_2$ give 1-bromo-1-TMS-2-hydroxy-pentane (*101*) whereas five equivalents of $MgBr_2$ directly lead to 1-TMS-2-pentanone (*102*). Another very interesting alternative is the application of

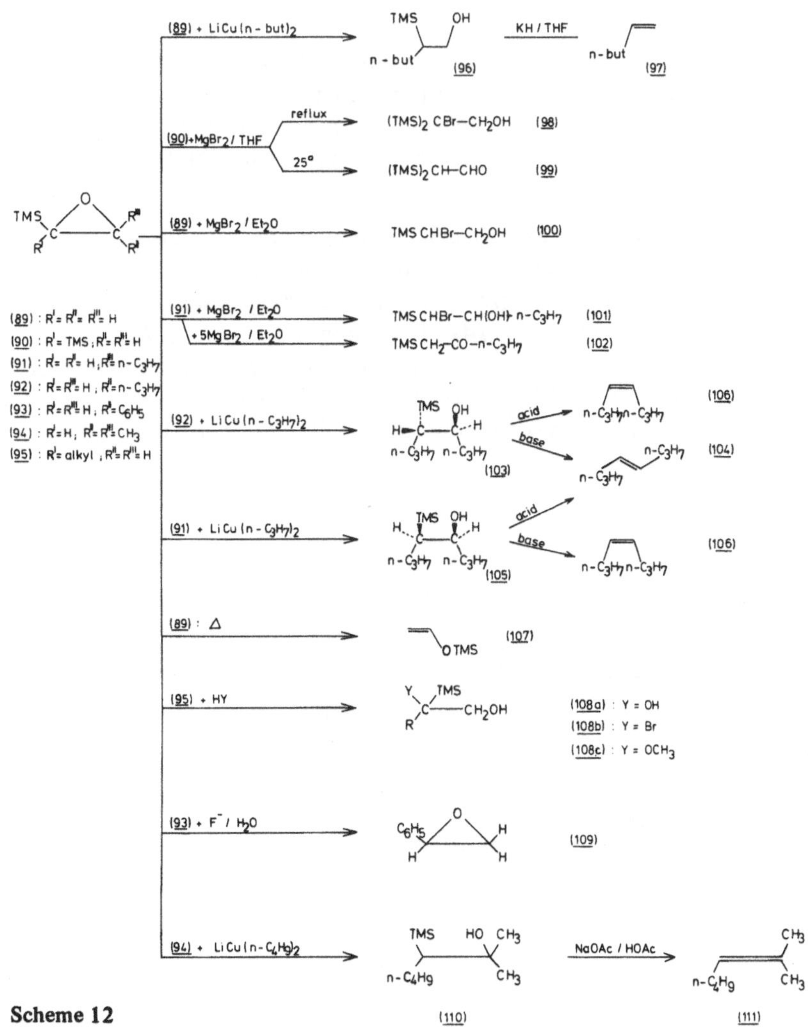

Scheme 12

organometallic reagents[75] such as diorganolithium cuprates. TMS-ethylenoxide (*89*) plus lithio-di-(n-butyl) cuprate afford 2-TMS-2-hexanol (*96*). Further treatment with potassium hydride in THF gives 1-hexene (*97*). Analogously, 1,1-dimethyl-2-TMS-oxirane (*94*) plus lithio(n-butyl) cuprate induces ring opening to 2-methyl-2-hydroxy-3-TMS-heptane (*110*) which is then converted to 1,1-dimethyl-1-hexene (*111*) using sodium acetate/acetic acid. In an equal manner Z-1-n-propyl-2-TMS-oxirane (*91*) and E-1-n-propyl-2-TMS-oxirane (*92*) react with lithio-di(n-propyl) cuprate and subsequent base treatment yields E-4-octene (*104*) and Z-4-octene (*106*), respectively, whereas acid catalysis affords the opposite products, i. e. E- leads to the Z- and Z- to the E-product, respectively[75]. Quite remarkable is the influence of temperature. Thus, 1,1-bis-TMS-oxirane (*90*) plus magnesium bromide in THF as solvent give at room temperature α,α-bis(TMS)-acetaldehyde (*99*), whereas on reflux 2-bromo-2,2-bis(TMS)-ethanol (*100*) can be isolated[76]. It could be shown that pyrolysis of TMS-ethylene oxide (*89*) opens the way to trimethylsiloxy-ethene (*107*)[77]. Ring opening of 1-alkyl-1-TMS-oxirane (*95*) with water, hydrobromic acid, methanol yields 2-hydroxy-(*108a*), 2-bromo- (*108b*), 2-methoxy-2-alkyl-2-TMS-ethanol (*108c*)[78], respectively. In addition to the above described procedures, another alternative has to be mentioned, the smooth displacement of the silyl moiety by means of fluoride ion catalysis, leaving the oxirane ring intact[79]. This method enables phenyl-oxirane (*109*) to be synthesized from E-1-phenyl-2-TMS-oxirane (*93*) (Scheme 12). For further ring opening reactions see also[80, 81].

The reactions of 1-TMS-cyclohexene oxide are similar (*112*). Treatment[78, 82] with sulfuric acid/water (in acetone), concentrated hydrobromic acid, sulfuric acid/methanol, lithium aluminum hydride afford the corresponding compounds 1,2-dihydroxy- (*113*)-, 1-bromo-2-hydroxy- (*114*)-, 1-methoxy-2-hydroxy- (*115*)-, 1-hydroxy-2-TMS-cyclohexane (*117*). Application of base to *115* yields 1-methoxy-1-cyclohexene (*116*). Pyrolysis of *112* gives a mixture of 1-trimethylsiloxy-1-cyclohexene (*118*) and 3-trimethylsiloxy-1-cyclohexene (*119*)[77] (Scheme 13).

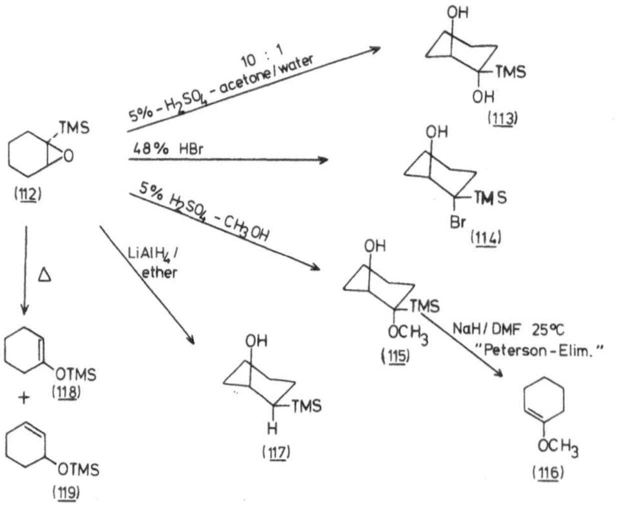

TMS
O
(112)
10 : 1
5% - H_2SO_4 - acetone / water
OH
TMS
OH
(113)
48% HBr
OH
TMS
Br
(114)
5% H_2SO_4 - CH_3OH
OH
TMS
OCH_3
(115)
NaH / DMF 25°C
"Peterson-Elim."
OCH_3
(116)
LiAlH$_4$ / ether
OH
TMS
H
(117)
Δ
OTMS
(118)
+
OTMS
(119)

Scheme 13

5 Hydrogenations by Means of Silanes

Besides the hydrosilylation already described (vide supra), silanes are able to hydrogenate a great variety of substrates. Since the early 1960's a common method in the peptide chemistry is the cleavage of the carbobenzoxy(cbo)protection group via Et_3SiH (*120*)/$PdCl_2$[83] which enables the amino acid to be obtained in a one step reaction (Scheme 14):

$$\underset{\substack{|\\ H-N-C(=O)-OCH_2C_6H_5}}{R-CH-COOH} \xrightarrow[(120)]{HSiEt_3/PdCl_2/CH_3OH} \underset{\substack{|\\ NH_2}}{R-CH-COOH} + Et_3SiOCH_3$$

Scheme 14

In the meantime a great number[84–92] of similar hydrogenations have been described which – instead of the above mentioned noble metal catalyst – mainly use strong acids (i. e. trifluoroacetic acid[88] and mineral acids[91]) or Lewis acids (i. e. aluminum chloride[89] and boron trifluoride[90]).

Very well studied are arylaliphatic compounds[84], especially those which have one double bond. By varying the ratio of carbonyl:silane:acid, the course of reaction can be immensely influenced. If the ratio is 1:1:10, only the double bond is hydrogenated[85] whereas a mixture 1:3:10 hydrogenates both the double bond and the carbonyl function[85]. When this reducing combination is used in the former ratio, 3-benzoyl-butyric acid yields 4-phenyl-valeric acid (*122*), p,p'-dinitrobenzophenone p,p'-dinitrodiphenylmethane (*123*), 1,6-diphenyl-hexane-1,6-di-one 1,6-diphenyl-hexane (*124*) and p-methoxybenzaldehyde p-methoxytoluene (*125*), respectively[84]. 1,3-Diphenyl-1-propene-3-one gives (depending on the ratio ketone:silane:acid) 1,3-diphenyl-1-propanone (*126*) (1:1:10) and 1,3-diphenyl-propane (*127*) (1:3:10), respectively[85].

Heterocyclic systems can also be reduced by this procedure. 2-benzoyl-thiophene and 3-methyl-thiophene undergo reduction to 2-benzyl-(*128*)- and 3-methyl-thiophane (*129*), respectively[86]. Analogously, 1-methyl-cyclohexane (*132*) can be obtained from 1-methyl-1-cyclohexene[88].

If aluminum chloride is used as catalyst, dehalogenation occurs. If cyclohexyl bromide is treated with Et_3SiH (*120*)/$AlCl_3$, cyclohexane (*130*) can be achieved. The combination R_3SiH/BF_3, on the other hand, enables hydrocarbons to be synthesize from their alcohols, for instance, 2-phenyl-3,3-dimethyl-2-butanol affords 2-phenyl-3,3-dimethyl-butane (*131*)[90].

An expansion of the procedures described above has been elaborated by West and co-workers[91] who used *120*/acetonitrile/aqueous mineral acid to synthesize a great variety of N-substituted-acetamides. In the case of benzophenone (*135*), N-(diphenylmethyl)-acetamide (*136*) is isolated.

Moreover, a Rosenmund reduction can be carried out too. Triethylsilane (*120*) in presence of palladium/carbon converts acetylchloride into acetaldehyde (*133*)[92].

If a very simple reagent such as trichlorosilane (*2*) is applied to phenylphosphinoxides (e. g. 9-phenyl-9-phospha-9,10-anthraquinone) the corresponding phenylphosphine[93, 94] [e. g., 9-phenyl-9-phospha-anthrone (*134*)] is obtained (see Scheme 15).

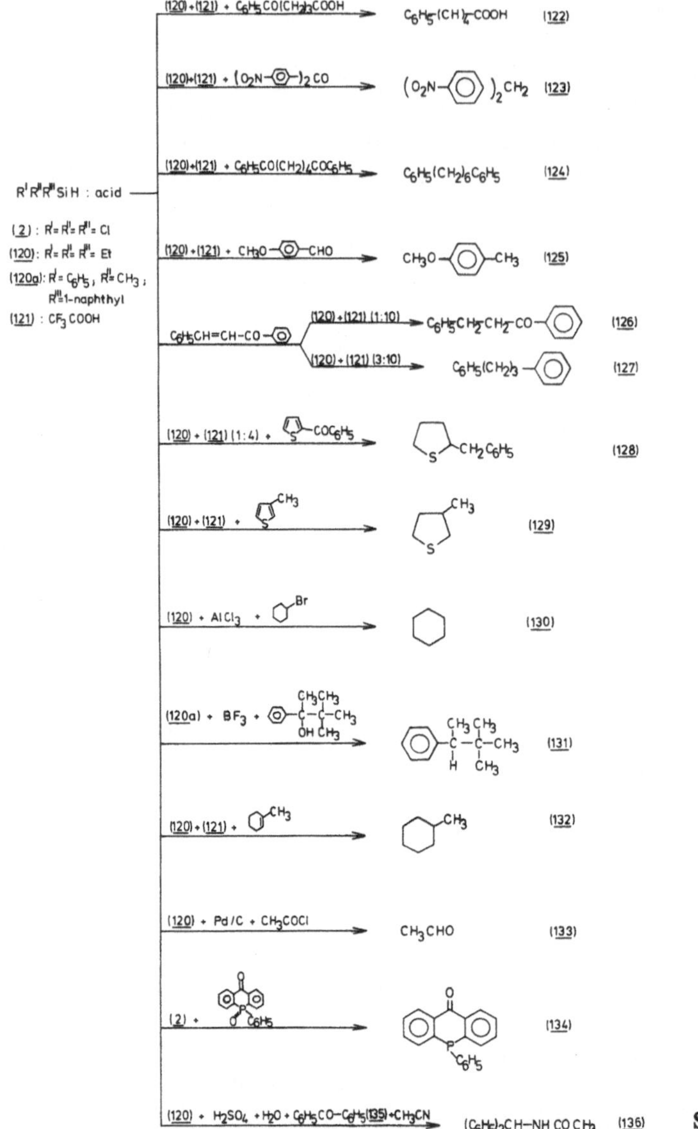

Scheme 15

Further applications of the silane/acid (i. e. *120*/trifluoroacetic acid) system include reductions of steroidal substrates[95, 96]. Another notable variant is the use of methylpolysiloxane (*137*)[88, 97] especially in presence of bis(dibutylacetoxytin)-oxide (DIBATO)[97], as hydrogenating agent; via this method benzophenone (*135*) yields diphenylmethanol (*138*) (Scheme 16). Under hydrosilylation-like conditions

$$\underset{(135)}{(C_6H_5)_2C{=}O} + \frac{1}{n}\underset{(137)}{\left[CH_3-\overset{\displaystyle O-}{\underset{|}{\overset{|}{Si}}}-H \right]_n} \xrightarrow{ROH/2\%\ DBATO} \underset{(138)}{(C_6H_5)_2C(H)-OH}$$

Scheme 16

(i. e. silane + noblemetal catalyst) *120* has great reducing potential, too. Because of its mild conditions, high yields and high selectivity, this reaction is very appropriate especially for the reduction of terpenoids[98]. Citral (*139*) is converted to the corresponding saturated aldehyde (*140*) (Scheme 17) after subsequent hydrolysis. This reaction can be transformed to imines, too[99, 100], yielding amines, especially with potential chirality if chiral catalysts are employed[100] (Scheme 18).

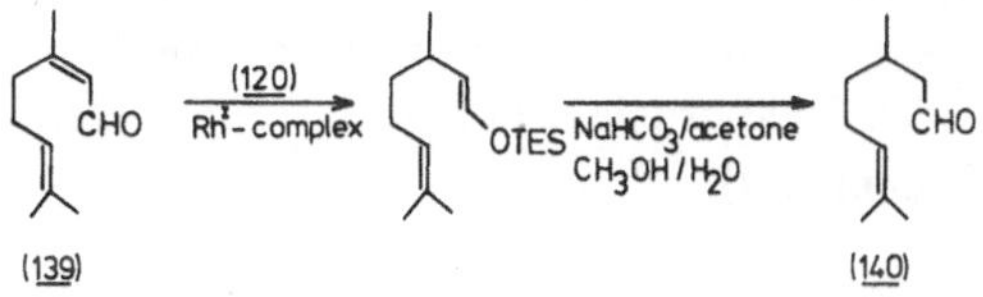

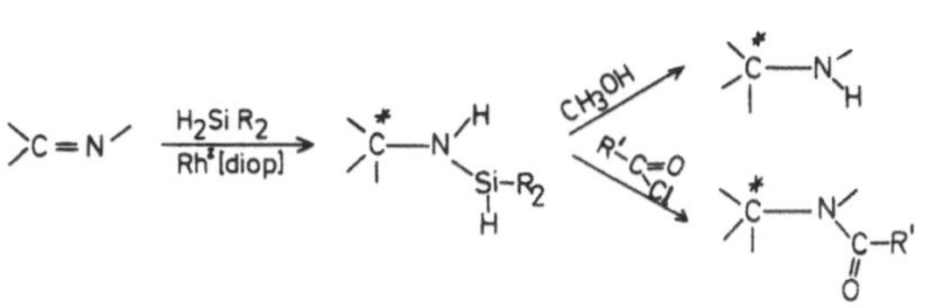

Scheme 17

Scheme 18

6 Reductive Silylation

The observations[101] that 1,2-bis(TMS)-1,2-dihydro-naphthalene (*141*)[101, 102] reacts with further chlorotrimethylsilane (*142*) to afford tris- and tetrakis(TMS)-tetrahydronaphthalenes (*143, 144a, b*) and rearrangements[101] were an early example of the so-called reductive silylation (Scheme 19).

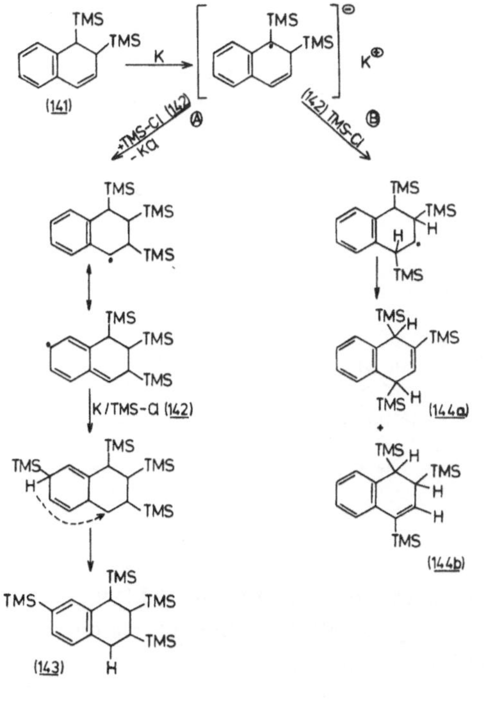

Scheme 19

Three products, 1,2,3,7-tetrakis(TMS)-1,2,3,4-tetrahydro-naphthalene (*143*), 1,2,4-tris(TMS)-1,4-dihydro (*144a*)- and 1,2,4-tris-(TMS)-1,2-dihydro-naphthalene (*144b*) were isolated[101].

The same mechanism is an explanation for the following results where 1-(trimethylsiloxy)naphthalene (*145*) and 1,5-bis(trimethylsiloxy)naphthalene (*146*) can be converted via reductive silylation into 1-TMS-naphthalene (*147*) and 1,5-bis(TMS)-naphthalene (*148*) respectively[103, 104] (Scheme 20).

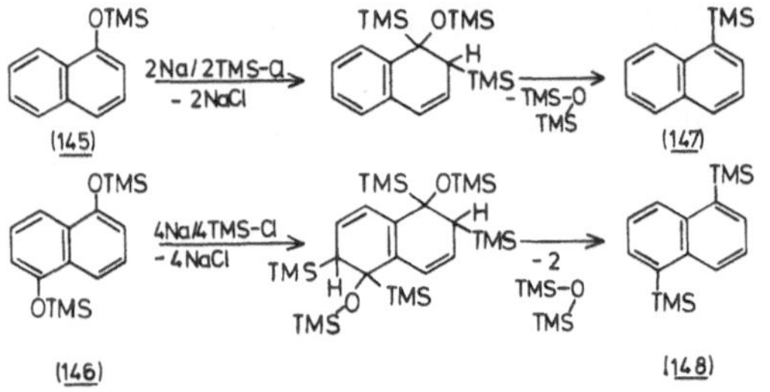

Scheme 20

The reductive silylation under these conditions renders possible an access to useful heterocyclic intermediates which undergo various electrophilic displacements (Scheme 21)[105] of the TMS moiety. In the first step phenazine (*149*) is converted to the corresponding 5,10-bis(TMS)-5,10-dihydrophenazine (*150*) that can be acylated in excellent yields to the desired 5,10-diacetyl-5,10-dihydro-phenazine (*151*).

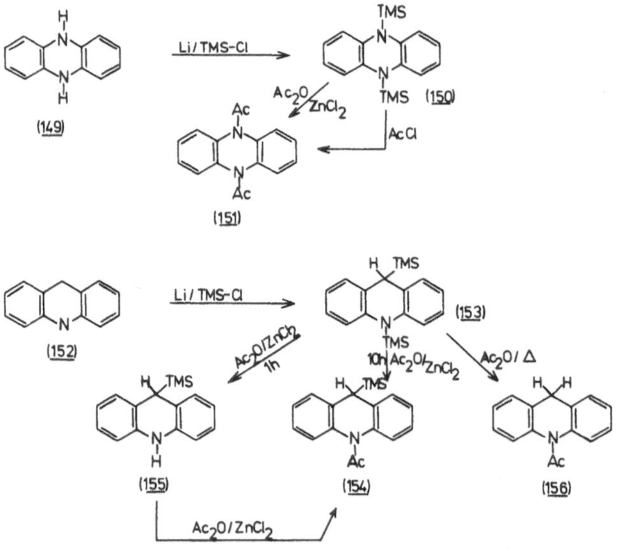

Scheme 21

Analogously, acridine (*152*) is silylated and the resulting 9,10-bis(TMS)-9,10-dihydroacridine (*153*) is then acylated to 10-acetyl-9-TMS-9-10-dihydroacridine (*154*) and 10-acetyl-9,10-di-hydroacridine (*156*), respectively[105].

If *153* plus the acylating agent is only heated briefly (1 h), *154* is not obtained, but 9-TMS-9,10-dihydroacridine (*155*) instead. Further treatment with the acylating agent leads to *154*. Recent developments, especially in France, revealed a broad spectrum of applications[14b)] using predominantly the two systems TMS-Cl (*142*)/Li/THF or *142*/Mg/HMPT.

Via the latter combination furane-2-(2-cyano-ethene-1-yl) (*160a*) and thiophene-2-(2-cyano-ethene-1-yl) (*160b*) are converted to furane-2-(2-cyano-1,2-bis(TMS)-ethane-1-yl) (*161a*) and thiophene-2-(2-cyano-1,2-bis(TMS)ethane-1-yl) (*161b*), res-

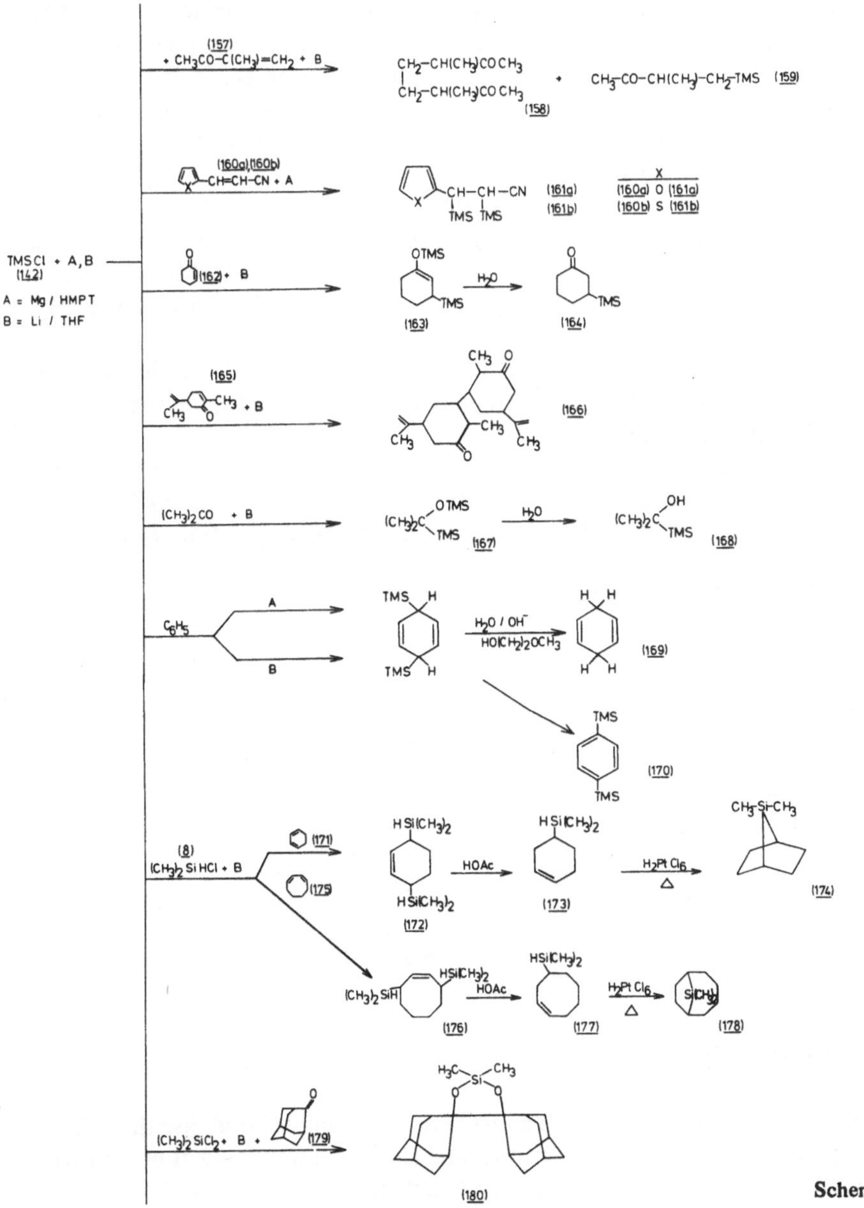

TMSCl + A,B (142)

A = Mg / HMPT

B = Li / THF

Scheme 22

pectively[106]. Using *142*/Li/THF, methyl-α-methylvinylketone (*157*) affords the dimeric 3,6-dimethyl-2,7-octadione (*158*) and the silylated 3-methyl-4-TMS-2-butanone (*159*)[107]. In terpenoid chemistry, carvone (*165*) yields under the same conditions the dimeric 6-1′,6′-dihydrocarvone-6′-yl)-1,6-dihydrocarvone (*166*)[107].

Similarly, cyclohexene-3-one (*162*) gives in the first step 1-trimethylsiloxy-3-TMS-1-cyclohexene (*163*) and by subsequent hydrolysis 3-TMS-cyclohexanone (*164*)[108]. If acetone is used instead, the system *142*/Li/THF leads to 2-trimethylsiloxy-2-TMS-propane (*167*) and after water treatment to 2-TMS-2-propanol (*168*)[108].

Via a very fascinating route, silicon-containing bicycles were synthesized[109]. The employment of system *B* (taking chlorodimethylsilane (8) instead of *142*) to 1,3-cyclohexadiene (*171*) and 1,3-cyclooctadiene (*175*) afford 3,6-dimethylsilylcyclohexene (*172*) and 3,8-dimethylsilylcyclooctene (*176*) respectively.

Partial desilylation by means of acetic acid gives the corresponding monosilyl products *173* and *177* and after subsequent intramolecular hydrosilylation 7,7-dimethyl-7-sila-bicyclo[2.2.1]-heptane (*174*) and 9,9-dimethyl-9-sila-bicyclo[4.2.1]-nonane (*178*), respectively, can be obtained[109]. Adamantanone (*179*) plus system *B* (using dichlorodimethylsilane instead of *142*) gives rise to 2,2-dimethyl-1,3-dioxa-2-sila-4,5-bis(adamanta)dispiro-cyclopentane (*180*)[110] (see also[111] for the alternative compound with two Me_2Si moieties. This reaction is affected via UV irradiation).

If benzene is treated under the conditions of A or B, either 1,4-cyclohexadiene (*169*) or 1,4-bis(TMS)benzene (*170*) can be synthesized depending on the reaction conditions[112] (Scheme 22).

Summarizing all these results, it can be said that three different kinds of products can be obtained due to reaction conditions[113]. The alternatives are: (a) "pure C-silylation"[113] under reduction; (b) dimerizing reduction ("duplication réductrice"[113]); (c) pure reduction ("simple réduction"[113]).

The scheme 23[113] illustrates these alternatives:

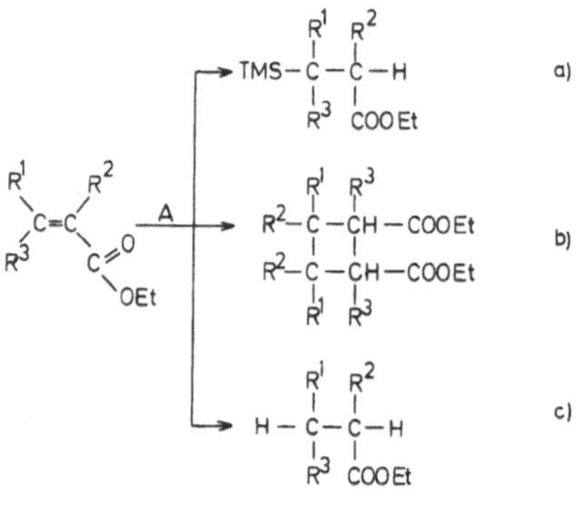

Scheme 23

C Reactions of Silanes with Carbonyl Compounds

Another class of compounds featuring highly reactive double bonds are those having a carbonyl function. Thus, mainly ketones and aldehydes were used as precursors to synthesize the universally applicable silyl enol ethers.

1 Silyl Enol Ethers

First attempts to achieve silyl enol ethers[114] are known since the late 1950's when hydrosilylation-type procedures were applied to α,β-unsaturated ketones[115–118] based upon the observation by Duffaut and Calas[119] that simple ketones are able to add trichlorosilane (*2*) under UV irradiation. These hydrosilylation reactions were widely expanded and intensively studied[120]. α,β-unsaturated ketones react via 1,4-addition[98] to silyl enol ethers[98] affecting only the conjugated double bonds. It is worth mentioning that the employment of chiral catalysts induces an asymmetrical reaction[25–27] (Scheme 24).

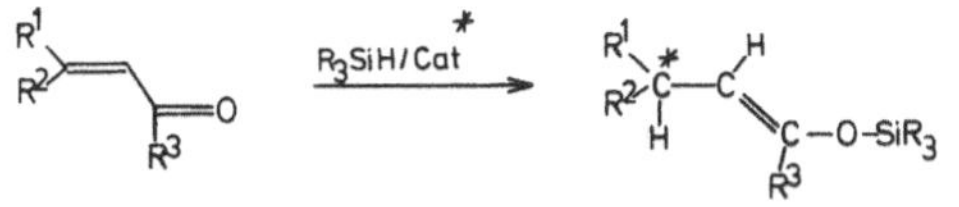

R_3SiH/Cat^*

Scheme 24

Besides the hydrosilylation, another type of reaction was developed using a strong base to form an enolate anion which is attacked by the added silylating agent (Scheme 25)[121–126].

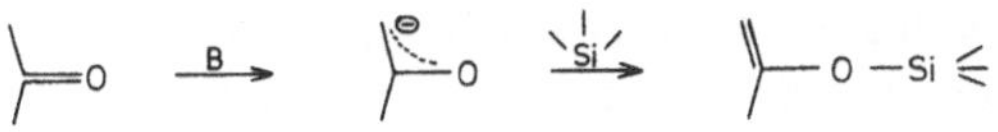

Scheme 25

More recently, the already described combinations A and B (chloro-organylsilane/metal/solvent; see section B 6: reductive silyation) proved[14b] to be more convenient because of its mild conditions, good reactivity and high yields (Scheme 26)[14b].

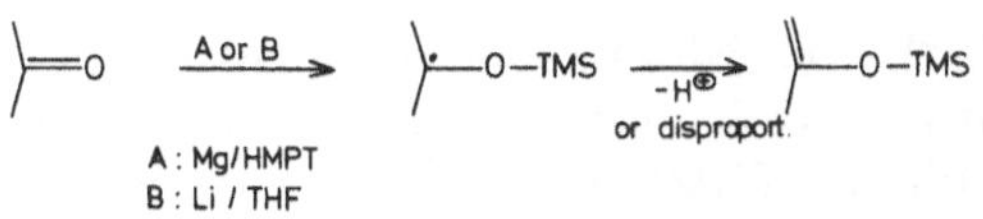

A or B

$-H^{\oplus}$ or disproport.

A : Mg/HMPT
B : Li / THF

Scheme 26

Moreover, the formation of enoxy-silanes via silylation of ketones[127] by means of N-methyl-N-TMS-acetamide (*172*) in presence of sodium trimethylsilanolate (*173*) was reported in 1969 and since then, the use of silylating reagents in presence of a catalyst has found wide appreciation and growing utilization as shown in recent papers[128–132] (Scheme 27). Diacetyl (*181*) can be converted by trifluoromethylsulfonic acid-TMS-ester (*182*) into 2,3-bis(trimethylsiloxy)-1,3-butadiene (*183*)[128] (for the application of *182* to nitriles, see[129]). Propiophenone (*184*) gives by treatment with ethyl TMS acetate (*185*)/tetrakis(n-butyl)amine fluoride 1-trimethylsiloxy-2-methyl-styrene (*186*)[130]. Cyclohexanone reacts with the combination dimethyl-TMS-amine (*187*)/p-toluenesulfonic acid to 1-trimethylsiloxy-1-cyclohexene (*188*)[131]. Similarly, acetylacetone plus phenyl-triethylsilyl-sulfide (*189*) afford 2-triethylsiloxy-2-pentene-4-one (*190*)[132].

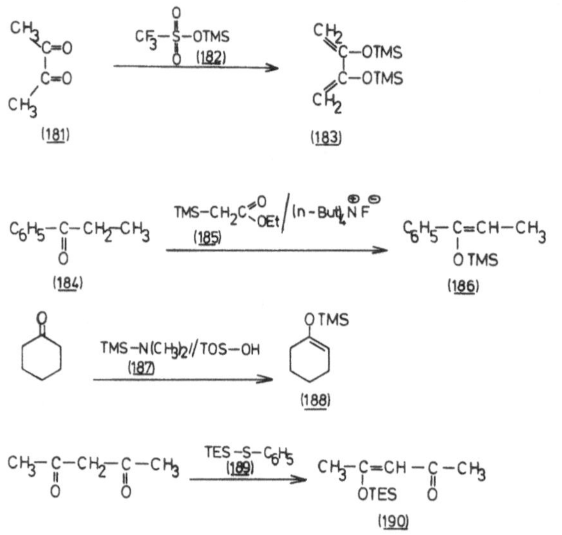

$CH_3COCOCH_3$ (181) $\xrightarrow{CF_3SO_2\text{-OTMS } (182)}$ $CH_2{=}C(OTMS){-}C(OTMS){=}CH_2$ (183)

$C_6H_5{-}CO{-}CH_2{-}CH_3$ (184) $\xrightarrow{TMS{-}CH_2COOEt\ (185)\ /\ (n\text{-}Bu)_4N^{\oplus}F^{\ominus}}$ $C_6H_5{-}C(OTMS){=}CH{-}CH_3$ (186)

Cyclohexanone $\xrightarrow{TMS{-}N(CH_3)_2\ (187)\ /\ TOS{-}OH}$ 1-OTMS-cyclohexene (188)

$CH_3{-}CO{-}CH_2{-}CO{-}CH_3$ $\xrightarrow{TES{-}S{-}C_6H_5\ (189)}$ $CH_3{-}C(OTES){=}CH{-}CO{-}CH_3$ (190)

Scheme 28

Because of their good reactivity and the good access, silyl enol ethers are important intermediates for a variety of products as shown below (Scheme 28).

2,2-Dimethyl-1-cyclohexyl-1-trimethylsiloxy-ethene (*191*) gives by means of methyllithium/ethylbromide in dimethoxyethane (DME) 1-cyclohexyl-2,2-dimethyl-1-butanone (*197*)[133]. Dimerizations occur when substances such as 1-trimethylsiloxy-styrene (*192*) or 1-trimethylsiloxy-1-cyclopentene (*195*) are treated with silver oxide/DMSO to afford 1,5-diphenyl-2,5-butadione (*198*) and 2,2′-dicyclopentanonyl (*199*)[134], respectively. Under the catalytic influence of Cu^{2+} ions, *192* plus benzenesulfonyl chloride yield phenyl-(1-phenyl-1-ethanone-2-yl)sulfone (*200*)[135].

The bromination of 1-phenyl-2-methyl-1-trimethylsiloxystyrene (*196*) leads to 1-phenyl-2-bromo-1-propanone (*201*)[136].

If 1-trimethylsiloxy-1-cyclohexene (*193*) is treated with borane/THF, hydrogen peroxide/alkali and then hydrolyzed, trans-1,2-cyclohexanediol (*203*) is obtained[137]; but trans-1-hydroxy-2-trimethylsiloxycyclohexane (*202*) can be isolated without subsequent acid-catalyzed hydrolysis[138] whereas the direct hydrolysis of the borane adduct *204* leads directly to cyclohexene (*205*)[139]. Very interesting is the use of $TiCl_4$ as catalyst. *193* plus benzaldehyde and $TiCl_4$ gives after hydrolysis 2-[hydroxy-(phenylmethyl)] cyclohexane-1-one (*206*)[140].

In a Michael-type addition *192* is converted by treatment with mesityl oxide (*207*) or methyl acrylate (*209*) under equal conditions ($TiCl_4$) to 1-phenyl-3,3-dimethyl-1,5-hexanedione (*208*) and methyl γ-benzoyl-butyrate (*210*)[141], respectively. *193* reacts with meta-chloroperbenzoic acid (*87*) to yield 2-trimethylsiloxy-cyclohexanone (*211*)[142].

Moreover, 3-benzoyl-2,4-pentanedione (*215*) is obtained when benzoylchloride is employed to the acetylacetone derivate *194*[143] whereas the reaction with benzaldehyde results a mixture of 3-benzylidene-2,4-pentanedione (*213*) and 3-[α-hydroxy-benzyl]-2,4-pentanedione (*214*)[143].

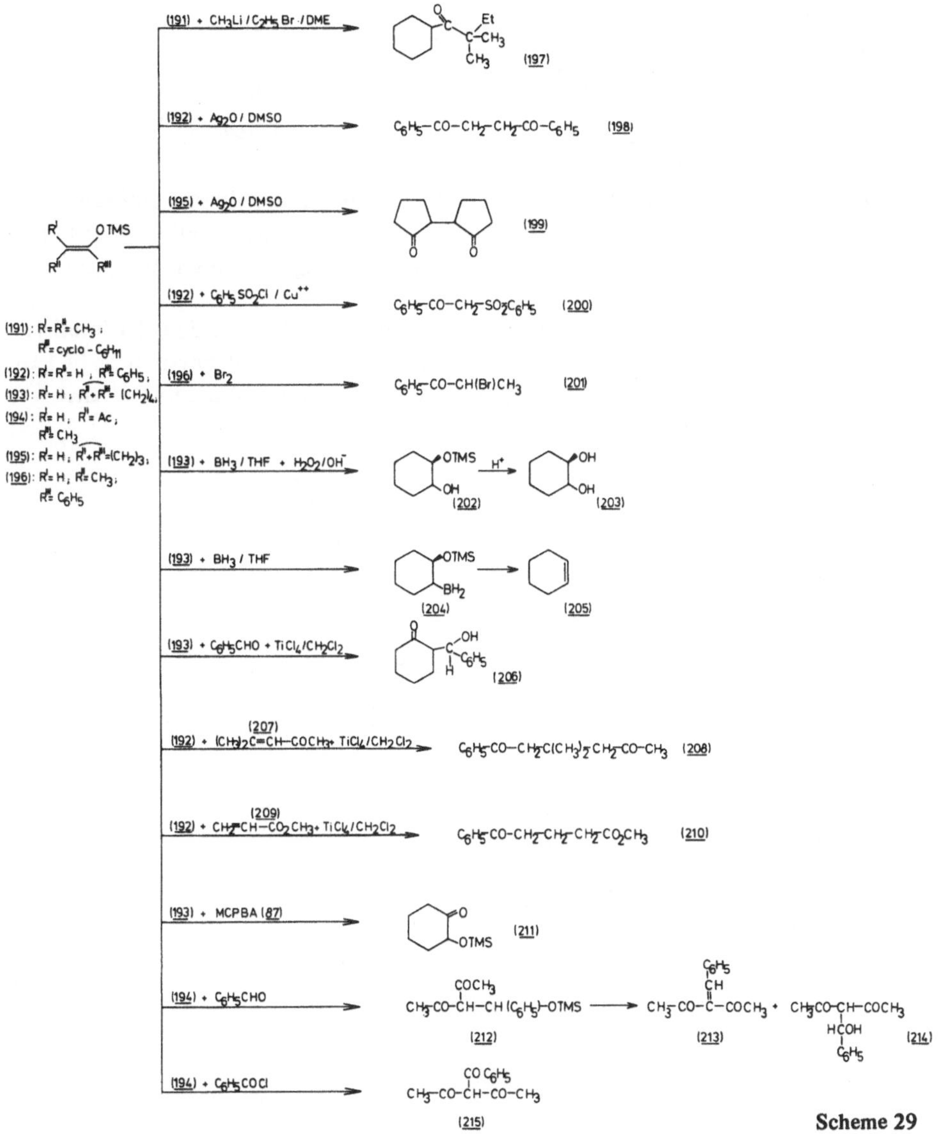

$R^I(R^{II})C{=}C(OTMS)R^{III}$

(191): $R^I = R^{II} = CH_3$; R^{III} = cyclo-C_6H_{11}
(192): $R^I = R^{II} = H$; $R^{III} = C_6H_5$;
(193): $R^I = H$; $R^{II} + R^{III} = (CH_2)_4$;
(194): $R^I = H$; $R^{II} = Ac$; $R^{III} = CH_3$
(195): $R^I = H$; $R^{II} + R^{III} = (CH_2)_3$;
(196): $R^I = H$; $R^{II} = CH_3$; $R^{III} = C_6H_5$

(191) + CH_3Li / C_2H_5Br / DME → (197)
(192) + Ag_2O / DMSO → $C_6H_5{-}CO{-}CH_2{-}CH_2{-}CO{-}C_6H_5$ (198)
(195) + Ag_2O / DMSO → (199)
(192) + $C_6H_5SO_2Cl$ / Cu^{++} → $C_6H_5{-}CO{-}CH_2{-}SO_2C_6H_5$ (200)
(196) + Br_2 → $C_6H_5{-}CO{-}CH(Br)CH_3$ (201)
(193) + BH_3 / THF + H_2O_2/OH^- → (202) $\xrightarrow{H^+}$ (203)
(193) + BH_3 / THF → (204) → (205)
(193) + C_6H_5CHO + $TiCl_4/CH_2Cl_2$ → (206)
(192) + $(CH_3)_2C{=}CH{-}COCH_3$ (207) + $TiCl_4/CH_2Cl_2$ → $C_6H_5{-}CO{-}CH_2C(CH_3)_2{-}CH_2{-}CO{-}CH_3$ (208)
(192) + $CH_2{=}CH{-}CO_2CH_3$ (209) + $TiCl_4/CH_2Cl_2$ → $C_6H_5{-}CO{-}CH_2{-}CH_2{-}CH_2{-}CO_2CH_3$ (210)
(193) + MCPBA (87) → (211)
(194) + C_6H_5CHO → $CH_3{-}CO{-}CH(COCH_3){-}CH(C_6H_5){-}OTMS$ (212) → $CH_3{-}CO{-}C({=}CH{-}C_6H_5){-}COCH_3$ (213) + $CH_3CO{-}CH(HCOH{-}C_6H_5){-}COCH_3$ (214)
(194) + C_6H_5COCl → $CH_3{-}CO{-}CH(COC_6H_5){-}CO{-}CH_3$ (215)

Scheme 29

2 Cyclopropanation Reactions

Conia and his co-workers[144] modified the Simmons-Smith reaction[145, 146] by using enol ether plus predominantly diiodomethane/zinc-silver couple instead of the common known diiodomethane/zinc-copper couple[147]. This method is an excellent access to cyclopropanols and related compounds[148, 149] (Scheme 29). Furthermore,

$$\text{enol-OTMS} \xrightarrow[\text{Zn-Ag}]{CH_2I_2 / \text{Zn-Cu}} \text{cyclopropyl-OTMS} \xrightarrow{H^{\oplus}} \text{cyclopropyl-OH}$$

(182) $\xrightarrow{CH_2I_2/Zn\text{-}Ag/Py}$ (216) $\xrightarrow{CH_3OH}$ (217)

Scheme 30

if 2,3-bis(trimethylsiloxy)-1,3-butadiene (*182*) is treated under these conditions, a twofold cyclopropanation is observed[150]. The formed 1,1'-bis(trimethylsiloxy)-bicyclopropyl (*216*) is then hydrolyzed to the corresponding 1,1'-dihydroxy-bicyclopropyl (*217*) (Scheme 30).

In the following, some applications are summarized (Scheme 31). 1-trimethylsiloxy-bicyclo[4.1.0]-heptane (*218*) gives after hydrolysis 1-hydroxy-bicyclo[4.1.0]-heptane (*255*)[151, 152] but further treatment with potassium t-butanolate/ether

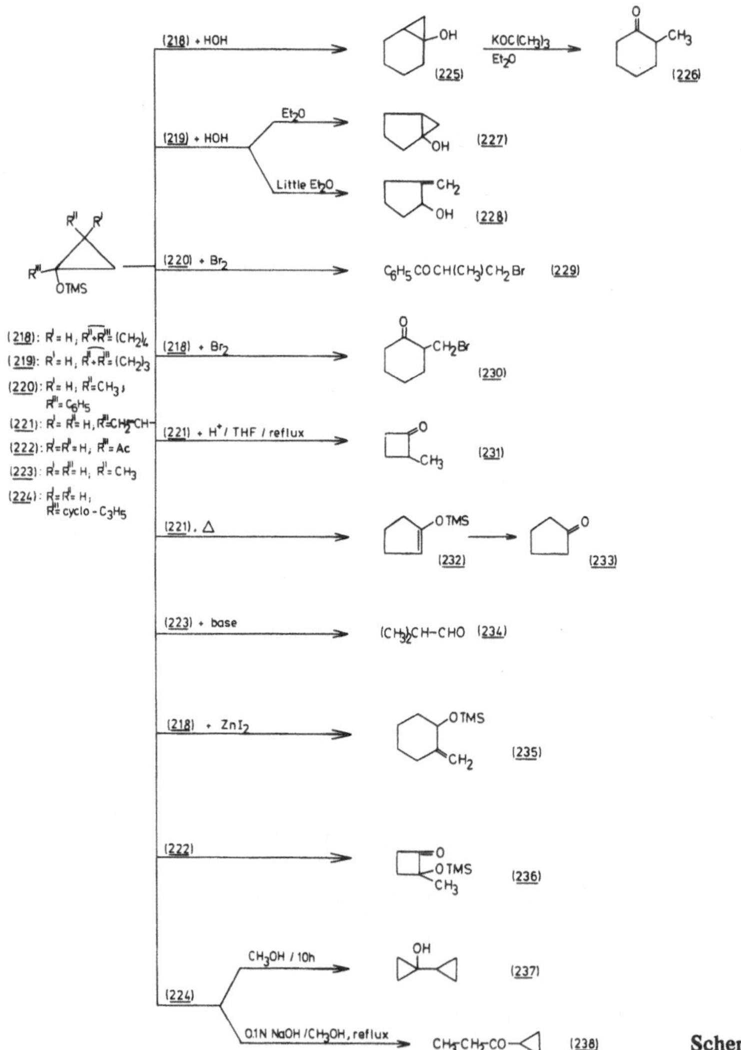

Scheme 31

yields 2-methyl-1-cyclohexanone (*226*)[152, 155, 156]. The cleavage of the cyclopropane ring shows a direct dependence on the dilution ratio[153] of the reaction mixture. In a not too concentrated medium (ether) the hydrolysis of 1-trimethylsiloxy-bicyclo[3.1.0]-hexane (*219*) affords 1-hydroxy-bycyclo[3.1.0]-hexane (*227*) whereas in a slightly diluted reaction mixture ring opening occurs and then 1-methylene-2-hydroxy-cyclopentane (*228*) can be isolated[153]. Another ring opening agent is bromine. The addition of bromine to *218* leads to 2-bromoethyl-1-cyclohexanone (*230*); similarly, starting from 1-phenyl-2-methyl-1-trimethylsiloxy-cyclopropane (*220*), 1-phenyl-2-methyl-3-bromo-1-propanone (*229*) is obtained[154]. Strong bases have been found to be mainly useful for indusing the opening of the cyclopropane ring affording α-methyl-carbonyl-compounds[152, 155, 156]; e. g. 1-methyl-1-trimethylsiloxy-cyclopropane (*223*) is converted to α-methyl-propanal (*234*)[155]. Because of its mild conditions and high selectivity, similar attempts were made in steroid chemistry[156].

Moreover, zinc iodide is able to function as a highly effective cleaving reagent, yielding 1-methylene-2-trimethylsiloxycyclohexane (*235*) if it is applied to *218*[157]. But it must be noted that zinc iodide is generated during the first reaction step thus it is imperative to separate the inorganic salt in order to avoid reactions leading to unwanted side effects[158].

Another important synthetic route are ring expansions, which can occur either by acid-catalyzed ring cleaving hydrolyzation or by pyrolysis[159]. Via these two alternative methods, 1-vinyl-1-trimethylsiloxy-cyclopropane (*221*) is transformed into 2-methyl-1-cyclo-butanone (*231*) or cyclopentanone (*232*), respectively[159]. By addition of pyridine, 1-acetyl-1-trimethylsiloxy-cyclopropane (*222*) is converted directly in the reaction mixture to 2-methyl-2-trimethylsiloxy-1-cyclobutanone (*236*)[160] (for further ring expansions see Conia and Robson[161]). Ring expansions were intensively studied by Trost and co-workers, too (e. g.: cf.[163]). One matter that concerns every preparative chemist is the hydrolysis. Methanol[162] as well as acid hydrolysis[158] predominantly gives rise to desilylation, e. g. 1-cyclopropyl-1-trimethylsiloxy-cyclopropane (*224*) affords under these conditions 1-cyclopropyl-1-cyclopropanol (*237*) whereas methanol in presence of small amounts of alkali transforms *224* under partial ring cleavage into 1-cyclopropyl-1-propanone (*238*).

3 Addition of Functional Organosilicon Compounds Across the Carbonyl Function

A very useful option for synthesis of α-functional siloxy compounds is the addition of variously substituted (functional) organosilanes. Both aldehydes and ketones react under these reactions although either basic or acidic catalysts are required in some cases (see Scheme 32).

If chloral (*239*) is treated with dimethyl-TMS-amine (*187*), 1-(trimethylsiloxy)-1-dimethylamino-2,2,2-trichloro-ethane (*244*) is obtained; via β-elimination this yields dimethylformamide (*245*)[164]. TMS-azide (*14*) forms with n-butanal (*240*) 1-azido-1-trimethyl-siloxy-butane (*246*)[165] [see section El].

In a similar reaction, ethyl α-TMS-acetate (*185*) and α-TMS-acetonitrile (*248*) plus benzaldehyde (*241*) affords ethyl 2-phenyl-2-trimethylsiloxy-propionate (*247*) and 2-phenyl-2-trimethylsiloxy-propionitrile (*249*), respectively[166].

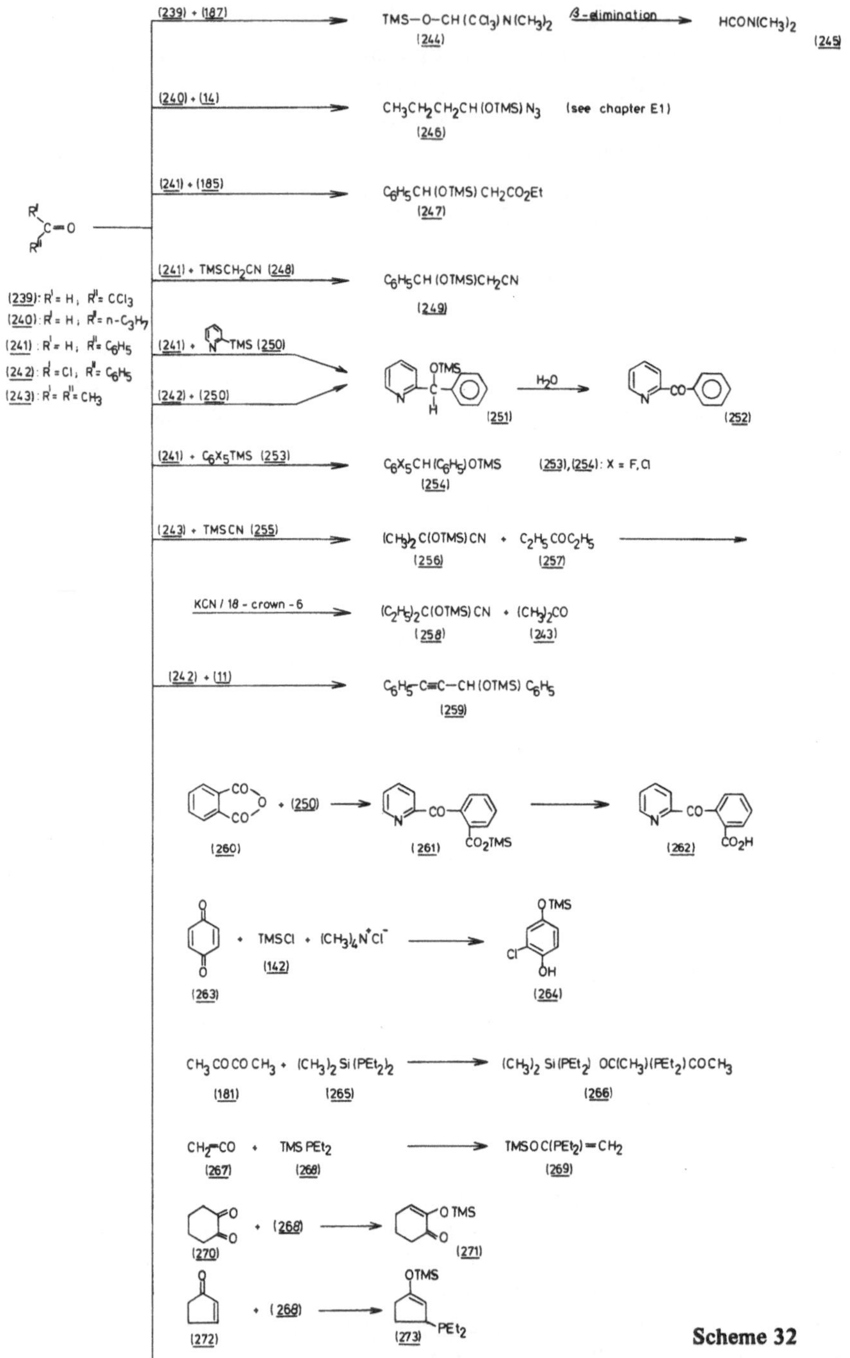

Scheme 32

These highly promising results were further modified and applied to the chemistry of heterocyclic compounds[167–169] as it is an easy pathway to prepare acylated heterocycles in a "one-pot reaction" which consists of the employment of a carbonyl component on the silylated heterocycle at higher temperatures. The reaction of 2-TMS-pyridine (*250*) with benzaldehyde (*241*)[167], or benzoyl chloride (*242*)[168] gives 2-(α-trimethylsiloxybenzyl) pyridine (*238*) and after subsequent hydrolysis 2-benzoyl-pyridine ("phenyl-pyridyl-ketone") (*252*) can be isolated (in the latter case[168], *252* is obtained directly via catalysis of the generated hydrogen chloride). Analogously, acid anhydrides show the same reactional behaviour[168]. If *250* is treated with phthalic anhydride (*260*), 2-(trimethylsiloxycarbonyl)phenyl-2-pyridyl-ketone (*261*) is formed which is then hydrolysed to the corresponding (2-hydroxy-carbonyl)-phenyl-2-pyridylketone (*262*). Perhalogeno-TMS-phenyl-compounds (*253*) and benzaldehyde (*241*) afford perhalogenophenyl-trimethylsiloxy-methanes (*254*)[169, 170], but a catalytic effect of the fluoride ion must be noted[170]. In an exchange reaction[171], α-methyl-α-trimethylsiloxy-propionitrile (*256*) transforms 3-pentanone (*257*) into α-ethyl-α-trimethylsiloxy-butyronitrile (*258*) by means of the KCN-18-crown-6-complex.

This reaction is a special modification of the cyanosilylation (vide infra).

Another class of carbonyl-containing substances are the quinones. Quinone (*263*) shows pure carbonyl behaviour by adding chlorotrimethylsilane (*142*) in presence of a quaternary ammonium salt to form 1-chloro-2-hydroxy-5-trimethylsiloxybenzene (*264*)[172] in a 1,4-addition-type reaction.

In the last decade silicon-containing organophosphorus compounds have found growing interest as reagents.

Especially, bis(diethylphosphino)dimethylsilane (*265*) and diethylphosphino-trimethylsilane (*268*) were investigated intensively[173–176].

It could be shown that diacetyl (*181*) and *265* give 3-[(diethylphosphino)di-methylsilyl]-3-diethylphosphino-2-butanone (*266*)[173]. If *268* applied to ketene (*267*), 1-diethylphosphino-1-trimethylsiloxy-ethene (*269*) is accomplished[174]. 1,2-cyclohexadione (*270*) and *268* cause a combined addition/elimination affording 2-trimethylsiloxy-1-cyclohexene-3-one (*271*) (!)[175] whereas 1-cyclopentene-3-one (*272*) and *268* undergo 1,4-addition to the corresponding 1-trimethylsiloxy-3-di-ethylphosphino-1-cyclopentene (*273*)[176]. Recent results by Evans and co-workers[177, 178] furnished evidence that many more sila-phosphorus compounds are suitable for similar additions across the >C=O double bond. Their investigations proved that trimethylsiloxyphosphines as well as phosphinic esters and amides are equally able to react in the pattern already described. The following common scheme (Scheme 33) may illustrate their findings:

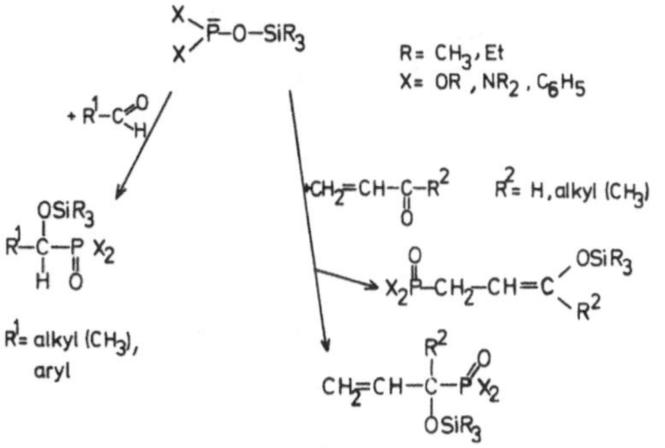

$X_2P{-}O{-}SiR_3$ R = CH_3, Et; X = OR, NR_2, C_6H_5

$+ R^1{-}CHO$ → $R^1{-}C(H)(OSiR_3){-}P(O)X_2$ (R^1 = alkyl (CH_3), aryl)

$+ CH_2{=}CH{-}C(O){-}R^2$ (R^2 = H, alkyl (CH_3)) → $X_2P(O){-}CH_2{-}CH{=}C(OSiR_3)R^2$ and $CH_2{=}CH{-}C(R^2)(OSiR_3){-}P(O)X_2$

Scheme 33

With saturated aldehydes, 1,2-addition is observed whereas unsaturated aldehydes and ketones give rise to a greater or lesser amount of 1,4-addition products. The ratio 1,2:1,4 is widely influenced by the kind of solvent used during the synthesis[178].

Concluding this part it must be stated that the addition of *11* across the carbonyl moiety in presence of quaternary ammonium fluorides has been achieved[179]. Benzaldehyde (*242*) and phenyl-TMS-acetylene (*11*) form (α-trimethylsiloxybenzyl)-phenylacetylene (*259*).

4 The Peterson Reaction

In 1968 Peterson published his results about the reaction of α-silylated carbanions with carbonyl compounds to achieve β-hydroxyalkyl-silanes. He found that instantaneous elimination affords unsaturated hydrocarbons ("Peterson olefination")[180]. Thus, the lithiated trimethyl-benzyl-silane (*274*) plus benzophenone (*135*) give via lithium 1,1,2,-triphenyl-2-TMS-ethanolate (*275*) 1,1,2-triphenyl-ethene (*276*)[180]

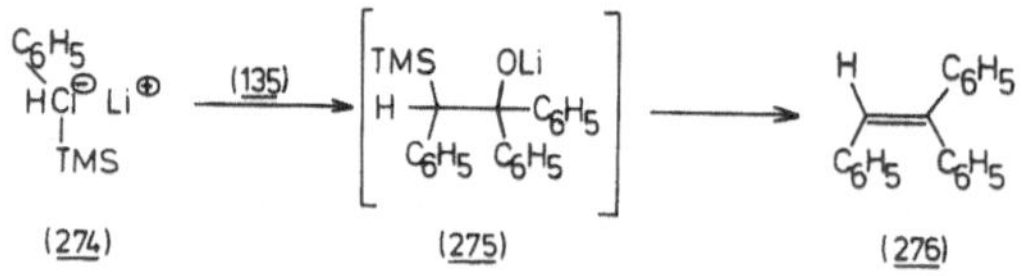

Scheme 34

(Scheme 34). Equally, the carbanions of bis- (*277*) and tris(TMS)-methane (*278*) and *135* form the corresponding single (*279*)[180, 181] and twofold silylated diphenylethene (*280*) (Scheme 35).

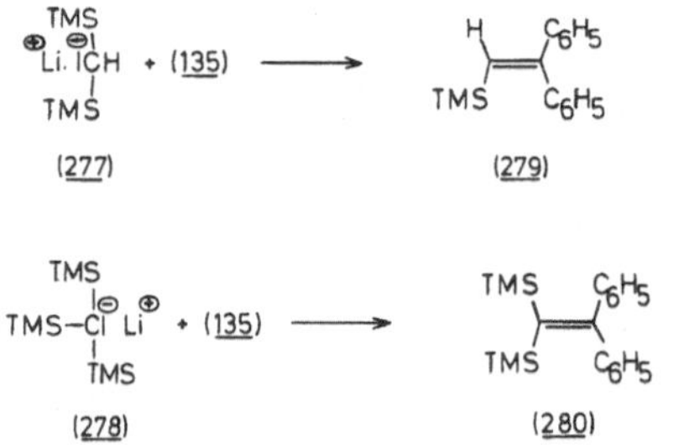

Scheme 35

Those β-hydroxyalkylsilanes resulting in the first reaction step (e.g. *275*) can be converted by means of thionyl[183] or acetyl chloride[184] into the corresponding chlorides which yield the desired alkenes via subsequent elimination under mild conditions (Scheme 36). The α-lithiated triphenyl-vinylsilane (*281*) and *135* give 3,3-di-

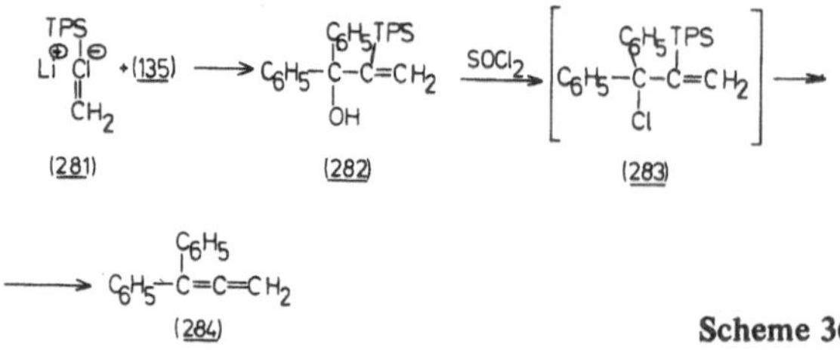

Scheme 36

phenyl-3-hydroxy-2-TPS-1-propene (*282*). Treatment with thionyl chloride leads to the chloride *283* which affords 1,1-diphenyl-allene (*284*)[183].

Another class of compounds which is easily obtained via this method are α,β-unsaturated esters when e.g. ethyl α-lithio-α-TMS-acetate (*285*) reacts with an aldehyde or a ketone[185–187] (Scheme 37). *285* plus acetaldehyde (*133*) give ethyl

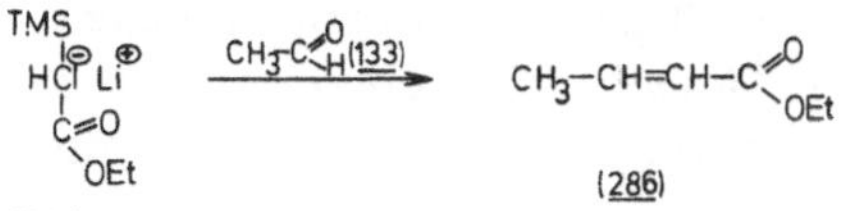

TMS–CH(Li)–C(=O)OEt (285) + CH_3–CHO (133) → CH_3–CH=CH–C(=O)OEt (286)

Scheme 37

crotonate (*286*). The course of reaction was intensively studied by Hudrlik and Peterson[188]. Their investigations showed that the stereochemistry of this elimination is strongly dependent on the kind of the applied reagent (for other olefin-forming reactions, see review[189]). In the following case (Scheme 38) the threo-alkane 5-TMS-4-octanol (*103a*) accomplishes the E-alkene *104* via syn elimination if treated with potassium hydride whereas the Z-alkene *106* is obtained in an anti-eliminating pathway when sodium acetate/acetic acid is applied.

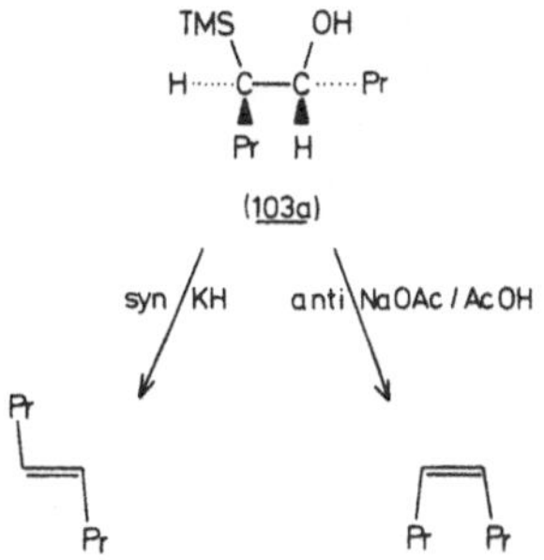

(103a) —syn/KH→ (104); (103a) —anti/NaOAc / AcOH→ (106)

Scheme 38

In a similar way, methylene moieties can be achieved by employing siliconeopentylmagnesium chloride (*287*) to carry out a "methylenation"[190]. If 3-pentanone (*257*) is treated with *287*, 2-ethyl-1-butene (*288*) is formed (Scheme 39). For further applications of the Peterson reaction and its modification, see also[64, 191, 192].

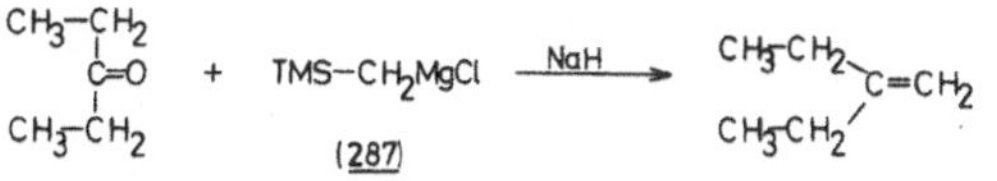

$(CH_3$–$CH_2)_2$C=O (257) + TMS–CH_2MgCl (287) —NaH→ $(CH_3$–$CH_2)_2$C=CH_2 (288)

Scheme 39

5 Reactions with Carboxylic Acid Esters[193]

Rühlmann and co-workers found[194–196] that carboxylic acid esters react under dimerisation to form acyloins if they are treated with sodium/TMS-Cl (*142*) in inert solvents. This Bouveault-Blanc-like synthesis has many parallels to the reductive silylation (vide supra).

The common scheme of the reduction is

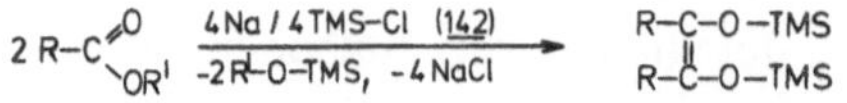

$$2\,R{-}C(=O)OR' \xrightarrow[-2R'{-}O{-}TMS,\ -4\,NaCl]{4\,Na\,/\,4\,TMS{-}Cl\ (142)} R{-}C(OTMS){=}C(OTMS){-}R$$

Scheme 40

This reaction offers some advantages over conventional methods as many acyloins can only be prepared by this method in good yields. Thus, by using ethyl acetate (*289*) in the first step 2,3-bis-(trimethylsiloxy)-2-butene (*290*) is obtained which can be easily hydrolysed to acetoin (*310*)[194]. Dicarboxylic acid esters undergo an intra-

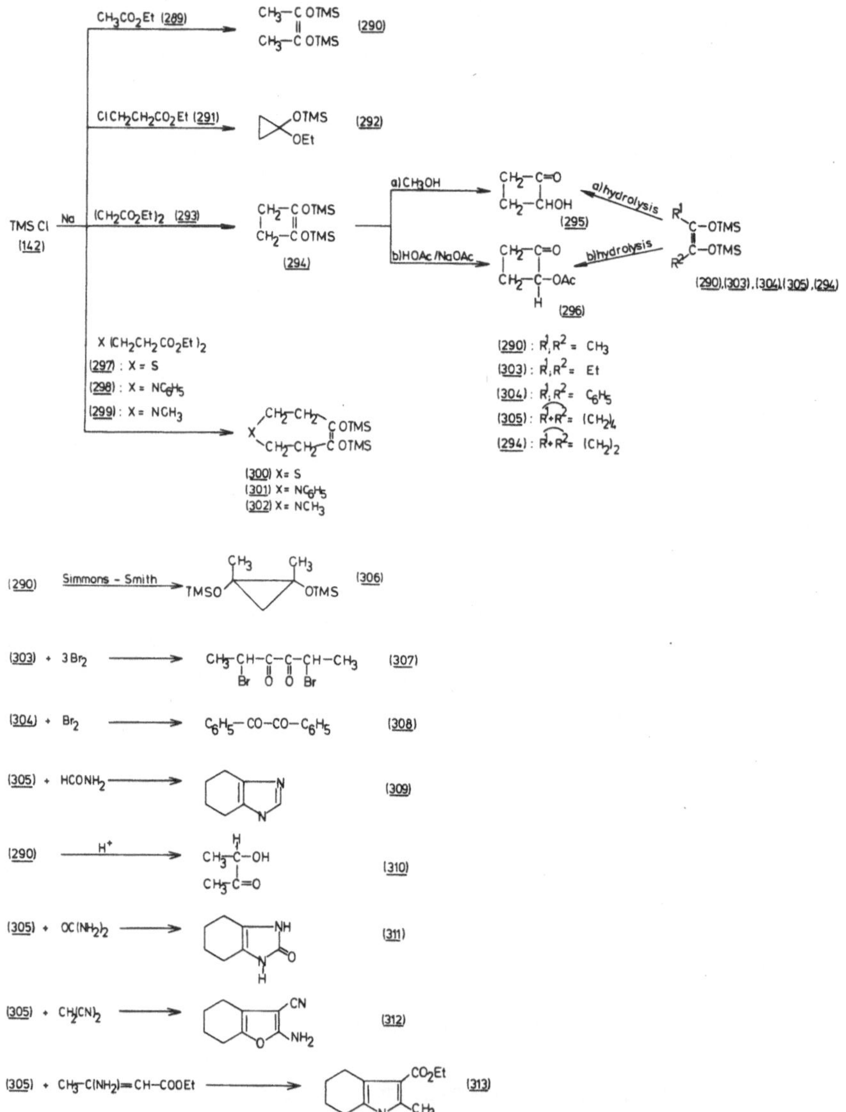

Scheme 41

molecular dimerisation if the reducing agent is employed[195, 196]; e.g. diethyl succinate (*293*) gives 1,2-bis(trimethylsiloxy)-1-cyclobutene (*294*) which can be hydrolysed either by methanol to form 2-hydroxy-1-cyclobutanone (*295*) or by employment of sodium acetate/acetic acid to 1-cyclobutanone-2-acetate (*296*)[196]. In an equal pattern bis(ethoxycarbonylethyl)sulfide (*297*), N-methyl- (*299*), N-phenyl-N,N-bis(ethoxy-carbonyl ethyl)amine (*298*) can cyclize to afford the heterocycles 4,5-bis(trimethylsiloxy)-2,3,6,7-tetrahydro-thiepine (*300*), 1-methyl- (*302*), 1-phenyl-4,5-bis(trimethylsiloxy)-perhydroazepine (*301*)[196].

Analogously, if applied to long chain dicarboxylic acid esters this cyclizing dimerisation yields macrocycles[195, 197] (Scheme 41). Although this reaction looks very simple and convenient, it must be stated that there is a strong dependence on the kind of solvent used during the first step[198]. In benzene or ether the Rühlmann reaction takes place yielding the bis(TMS)-enediolate (e.g. *315*) whereas in THF or dimethoxyethane (DME) a reductive silylation is observed, e.g. ethyl γ-methylbutyrate (*314*) gives 3-methyl-1,1-bis(TMS)-1-trimethylsiloxy-butane (*316*) (Scheme 42).

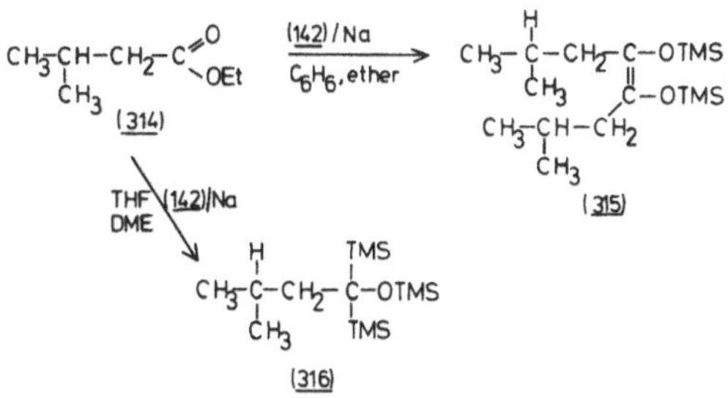

Scheme 42

With only two equivalents of *142*/Na employed, ethyl β-chloropropionate (*291*) is transformed after subsequent addition of *142* into 1-ethoxy-1-trimethylsiloxy-cyclopropane (*292*)[193].

Those bis(trimethylsiloxy)alkenes resulting in the first step are reactive intermediates which can be converted into a variety of compounds – they give an easy access to many products in a "one pot" synthesis. For instance via bromination of 3,4-bis(trimethylsiloxy)-3-hexene (*303*) 2,5-dibromo-3,4-hexadione (*307*) can be achieved[199] whereas 1,2-bis(trimethylsiloxy)stilbene (*304*) yields benzil (*308*)[199]. "Simmons-Smith conditions" were applied, too, and 2,3-bis(trimethylsiloxy)-2-butene (*290*) gives 1,2-dimethyl-1,2-bis(trimethylsiloxy)cyclopropane (*306*)[200].

Analogously to conventional reactions[145, 146, 201–203], bis(trimethylsiloxy) alkenes form heterocycles of different classes[193]. 1,2-bis(trimethylsiloxy)-1-cyclohexene (*305*) reacts with formamide[201], urea[202], malodinitrile[145, 146] and ethyl β-amino-crotonate[203] to afford 4,5,6,7-tetrahydrobenzimidazole (*309*)[193, 201], 2-oxo-1,3,4,5,6,7-hexahydrobenzimidazole (*311*)[193], 2-amino-3-cyano-4,5,6,7-tetrahydro-coumarone (*312*)[193] and 2-methyl-3-ethoxy-carbonyl-4,5,6,7-tetrahydro-indole (*313*)[193, 203], respectively.

D Synthesis of Silylated Heterocycles

1 Ring Formations

a) Pyrazoles

The reactions of silylated synthons to form heterocycles are well appreciated methods in the heterocyclic chemistry. A great variety of nitrogen containing products can be obtained in a classical approach via a 1,3-dipolarophilic addition of substituted acetylenes plus diazo-compounds[204] (Scheme 43). It could be shown[205] that these additions can be considered as evidence for a directive influence of the TMS moiety because one of the two possible reaction products is usually favoured (Scheme 43) and e.g. *320* is the only reaction product, whereas the addition of TMS-acetyl-acetylene (*317*) plus diazomethane (*318*) leads to two isomers featuring *325* as the main product (Scheme 43).

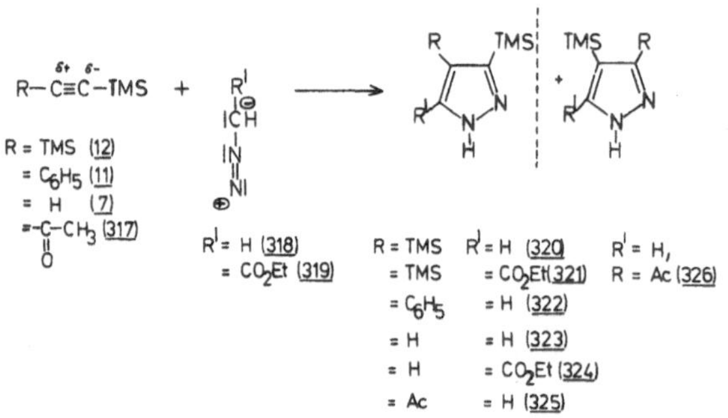

R = TMS (12); = C_6H_5 (11); = H (7); = $-C(=O)-CH_3$ (317)

R' = H (318); = CO_2Et (319)

R = TMS, R' = H (320); = TMS, = CO_2Et (321); = C_6H_5, = H (322); = H, = H (323); = H, = CO_2Et (324); = Ac, = H (325)

R' = H, R = Ac (326)

Scheme 43

As acetylene components, bis-TMS-acetylene (*12*), phenyl-TMS-acetylene (*11*), mono- TMS-acetylene (*7*) and TMS-acetyl-acetylene (*317*) were applied to diazomethane (*318*) and ethyl diazoacetate (*319*) giving the products *320–325* and *326*, respectively. If *320* is treated with bromine, the corresponding monobromo- (*328*) and di-

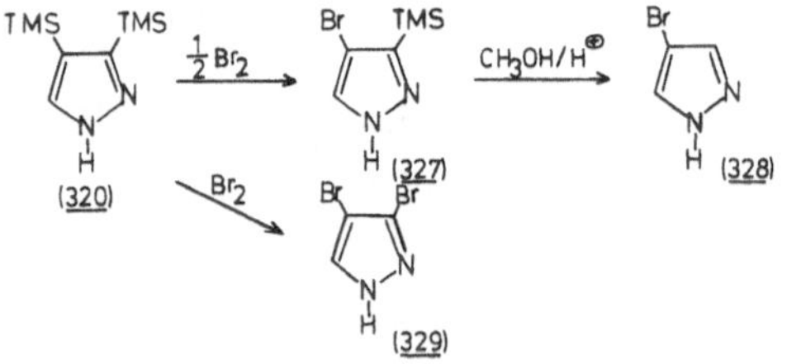

Scheme 44

bromo-pyrazole (*329*) can be isolated (Scheme 44)[204]. The application of nitrosating agents to *320, 323, 330* affords the nitrosopyrazoles *331, 333, 334* which can be readily oxidized to the corresponding nitro compounds *332, 335, 336* (Scheme 45), respectively[206].

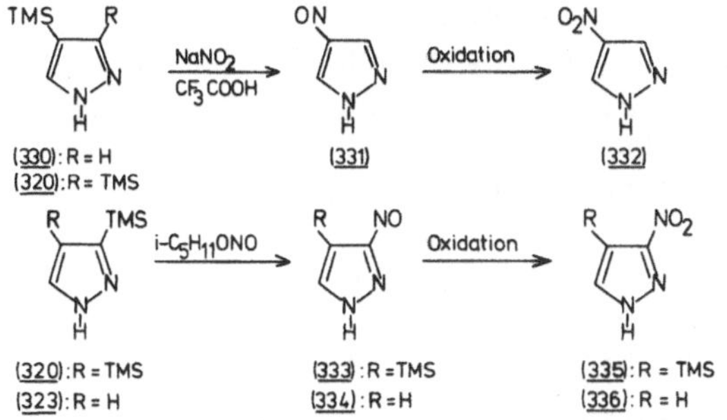

(330): R = H
(320): R = TMS
(331)
(332)
(320): R = TMS
(323): R = H
(333): R = TMS
(334): R = H
(335): R = TMS
(336): R = H

Scheme 45

The reaction of 3(5),4-bis(TMS)-pyrazole (*320*) with concentrated nitric acid or hydrogen chloride leads to 3(5),4-bis(TMS)-pyrazolium nitrate (*337*) or-chloride (*338*), respectively[207]. Thermolysis of *337* gives the hitherto unknown pure TMS-nitrate (*339*) in nearly quantitative yield (Scheme 46).

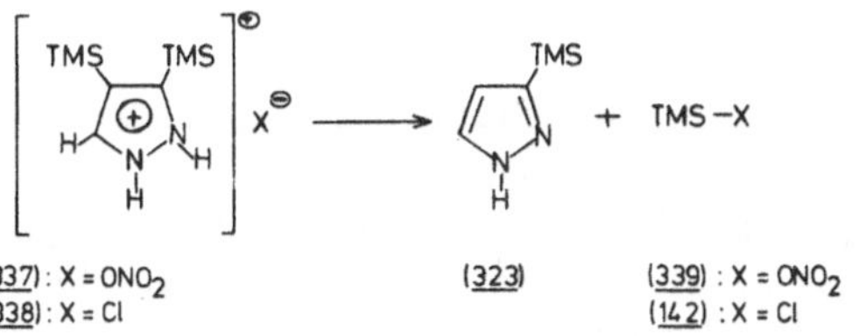

(337): X = ONO_2
(338): X = Cl
(323)
(339): X = ONO_2
(142): X = Cl

Scheme 46

The desilylation of C-silylpyrazoles can be achieved selectively: the replacement of silyl groups in 4-positions by hydrogen is smoothly effected by concentrated sulfuric acid. If concentrated nitric acid is used instead, 4-nitro-3(5)-TMS-pyrazole (*344*) can be isolated (Scheme 47)[205].

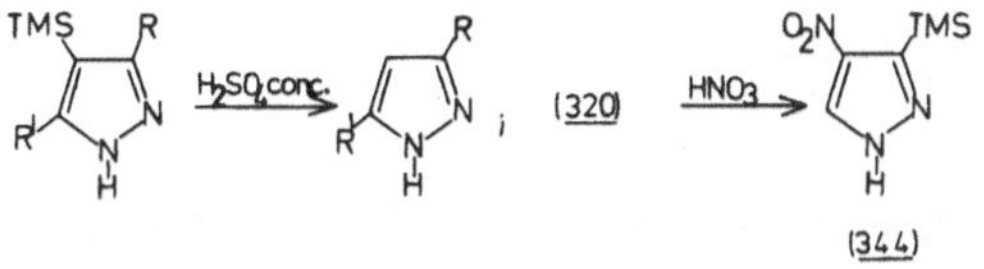

	R	R'	
(320)	TMS	H	(340)
(321)	TMS	CO_2Et	(341)
(326)	Ac	H	(342)
(323)	H	H	(343)
(335)	NO_2	H	(336)

(344)

Scheme 47

In contrast to these results, silyl moieties in 3(5)-positions are removed by common nucleophilic agents (e.g. C_2H_5ONa, $HO^{\ominus}$)[205] (Scheme 48).

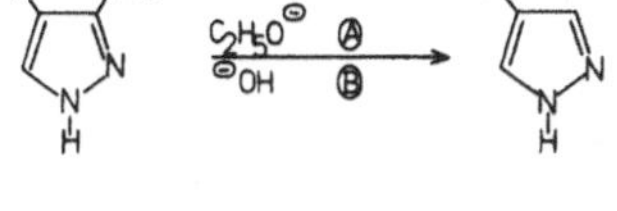

	R		Reag.
(320)	TMS	(330)	A
(323)	H	(343)	B
(322)	C_6H_5	(345)	B
(325)	Ac	(346)	B
(327)	Br	(328)	B
(344)	NO_2	(332)	B

Scheme 48

In an equivalent pattern, 3-phenyl-5-ethynyl-pyrazole (*349*) is synthesized by transforming 1,4-bis(TMS)-butadiyne (*38*) via benzoylchloride (*242*) into the corresponding 1-benzoyl-4-TMS-1,3-butadiyne (*347*)[48]. Subsequent acid-catalyzed ring closure with hydrazine hydrate and then hydrolysis under basic conditions yields 3-phenyl-5-ethynyl-pyrazole (*349*)[208] (Scheme 49).

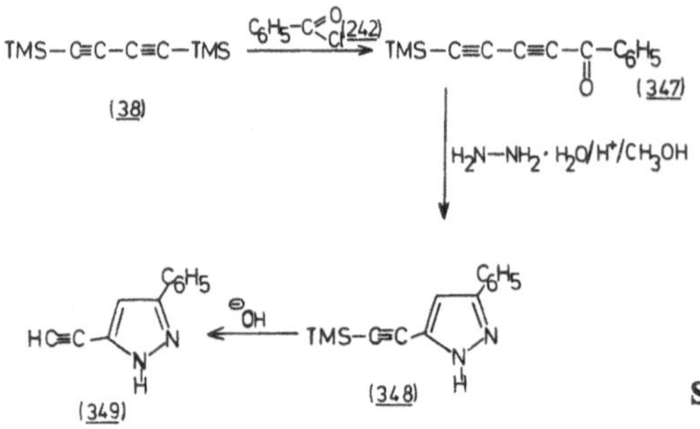

Scheme 49

b) Triazoles

Another class of heterocyclic compounds are silyl-substituted triazoles which can be obtained by the addition of TMS-azide (*14*) and manifold substituted acetylenes *1*, *35*, *350* (Scheme 50)[33, 209–211]. Those in the first step resulting 1-TMS-4-phenyl (*351*)-, 1-TMS-4,5-bis(methoxycarbonyl)- (*352*) and 1-TMS-4-alkyl-1,2,3-triazole (*353*) are then hydrolyzed to form 4-phenyl- (*354*)-, 4,5-bis(methoxycarbonyl)- (*355*), 4-alkyl-1,2,3-triazole (*356*).

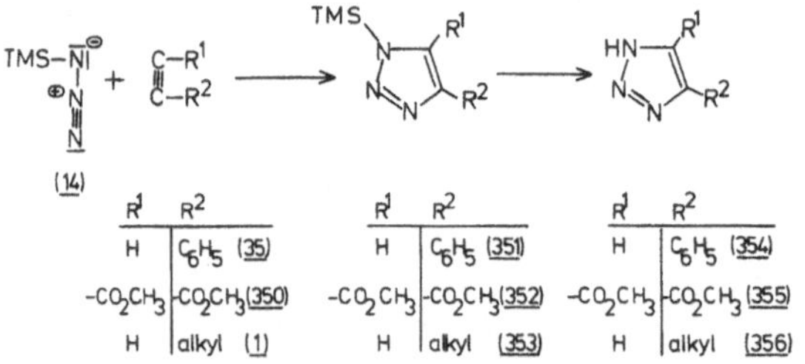

R^1	R^2
H	C_6H_5 (35)
$-CO_2CH_3$	$-CO_2CH_3$ (350)
H	alkyl (1)

R^1	R^2
H	C_6H_5 (351)
$-CO_2CH_3$	$-CO_2CH_3$ (352)
H	alkyl (353)

R^1	R^2
H	C_6H_5 (354)
$-CO_2CH_3$	$-CO_2CH_3$ (355)
H	alkyl (356)

Scheme 50

In a further reaction 2-TMS-1,2,3-triazole (synthesized by silylation of a 1,2,3-triazole[211]) can easily be acylated in 1-position using acetic anhydride or acetyl chloride[211] (Scheme 51). Tanaka et al.[212] have intensively studied this reaction using (p-subst)-monophenyl-acetylenes and *14*. Their results showed that in this case an electron releasing substituent induces a higher yield whereas it can normally be said: the more electron withdrawing a substituent is, the higher the yield in this type of reaction.

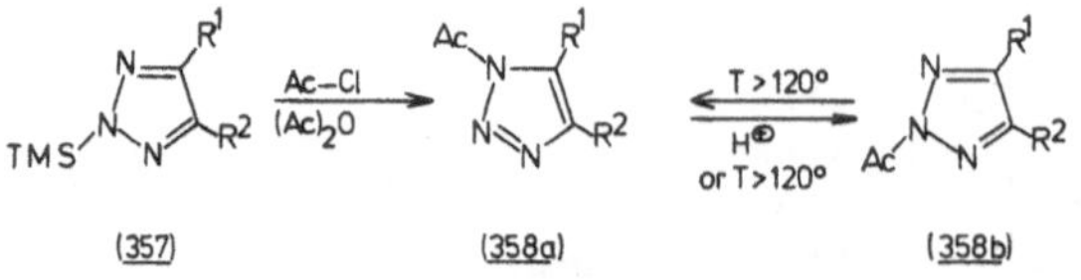

Scheme 51

c) Isoxazoles

Another very elegant reaction is the synthesis of carbon-silylated isoxazoles[213] by means of a 1,3-dipolarophilic cycloaddition of nitrile oxide with silylated acetylenes (*7*), (*12*) (Scheme 52). If mesitylnitrile oxide (*361*) and 3,3-dimethyl-3-sila-1,4-pen-

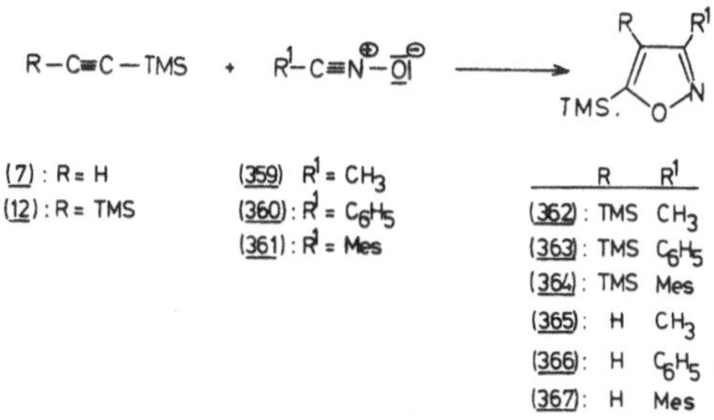

R—C≡C—TMS + R^1—C≡N$^{\oplus}$—O$^{\ominus}$ ⟶

(7) : R = H
(12) : R = TMS

(359) R^1 = CH_3
(360) : R^1 = C_6H_5
(361) : R^1 = Mes

	R	R^1
(362) :	TMS	CH_3
(363) :	TMS	C_6H_5
(364) :	TMS	Mes
(365) :	H	CH_3
(366) :	H	C_6H_5
(367) :	H	Mes

Scheme 52

tadiyne (*368*) or 3,3,4,4,-tetramethyl-3,4-disila-1,5-hexadiyne (*369*), respectively, are taken instead, one is able to synthesize the dimeric bis-(3-mesityl-5-isoxazolyl)-silanes *370* and *371*, respectively[213] (Scheme 53). Treatment of *362* with bromine

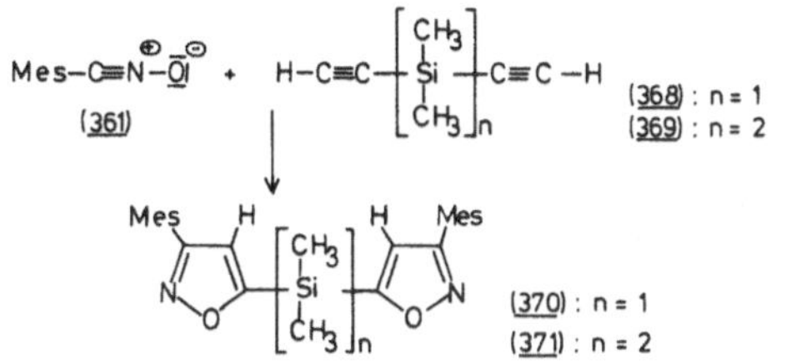

Mes—C≡N$^{\oplus}$—O$^{\ominus}$ (361) + H—C≡C—[Si(CH_3)$_2$]$_n$—C≡C—H

(368) : n = 1
(369) : n = 2

(370) : n = 1
(371) : n = 2

Scheme 53

gives under partial desilylation 4-bromo-3-methyl-5-TMS-isoxazole (*372*) whereas the employment of sulfuric acid affords 3-methyl-5-TMS-isoxazole (*365*) (Scheme 54)[213].

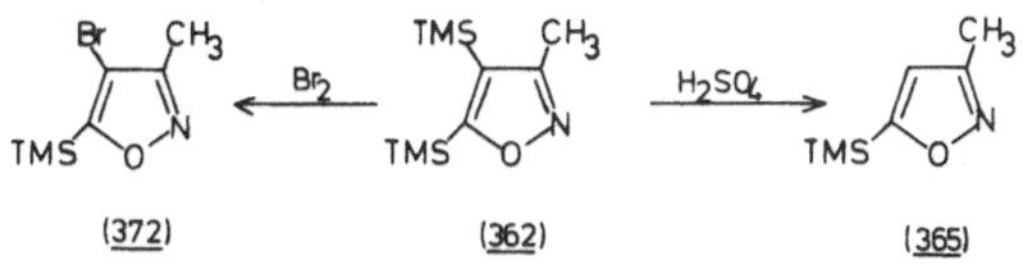

(372) ⟵ Br_2 — (362) — H_2SO_4 ⟶ (365)

Scheme 54

An interesting extension of these ring forming reactions has been reported by Washburne and co-workers[214] by using nitriles, e.g. benzonitrile (*373*).

In the first step the formation of 5-phenyl-2-TMS-1,2,3,4,-tetrazole (*374*) occurs which can either be hydrolyzed to 5-phenyl-1,2,3,4-tetrazole (*375*) or pyrolysed to give a N-TMS-benzaldehydehydrazonium compound (*376*). *376* can furthermore either dimerize to form 3,6-diphenyl-1,2,5-bis(TMS)-2,5-dihydro-1,2,4,5-tetrazine (*377*) and after subsequent hydrolysis and oxidation 3,6-diphenyl-1,2,4,5-tetrazine (*379*) or on the other hand react with a further equivalent *373* yielding in the last step 3,5-diphenyl-1,2,4-triazole (*381*)[214] (Scheme 55).

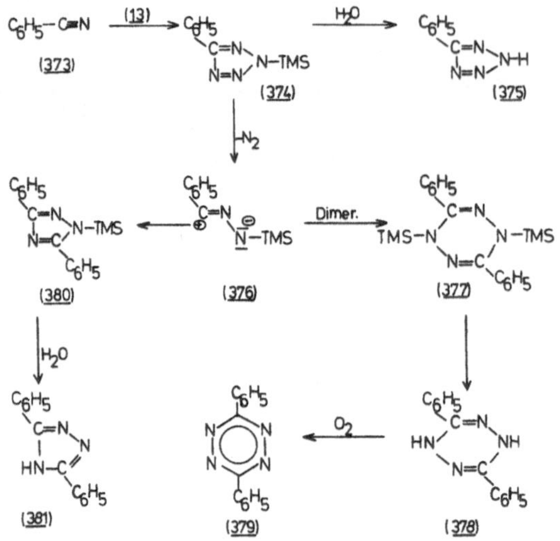

$C_6H_5-C{\equiv}N$ (373) —(13)→ (374) —H_2O→ (375)

(374) —$-N_2$→ (376) —Dimer.→ (377) → (378) —O_2→ (379)

(376) → (380) —H_2O→ (381)

Scheme 55

Equally, the synthesis of 5-(ferrocenyl)-tetrazole (*384*) could be achieved, too[214] (Scheme 56).

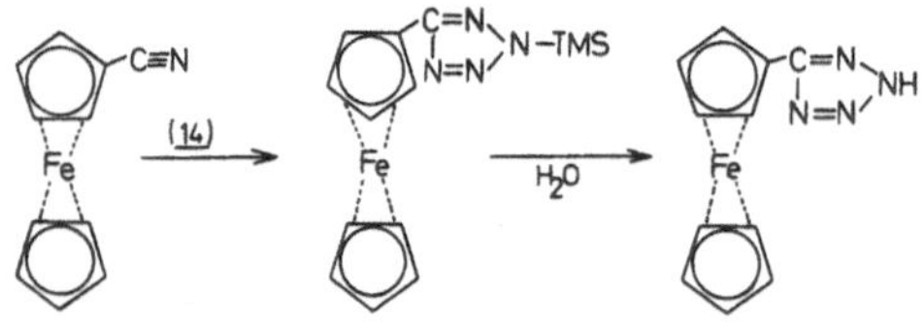

(382) —(14)→ (383) —H_2O→ (384)

Scheme 56

d) 1,3,4-Oxadiazolines

Kricheldorf could show that 5-substituted 1,3,4-oxadiazoline-2-one (*388–390*) can be formed via silylated intermediates of phenyl-3-acylcarbazates (*385–387*)[215] (Scheme 57).

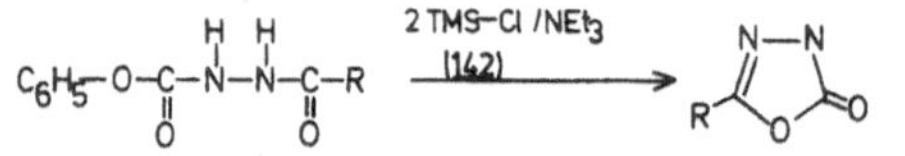

$C_6H_5-O-C(=O)-NH-NH-C(=O)-R$ —2 TMS–Cl /NEt_3 (142)→

(385) : R = CH_3
(386) : R = C_6H_5
(387) : R = t-But

(388) : R = CH_3
(389) : R = C_6H_5
(390) : R = t-But

Scheme 57

e) Pyridines

A very promising procedure deserves to be mentioned, the reaction of carbon suboxide (*392*) with hexamethyldisilazane (*393*)[216] which affords in the first step 2,4-dioxo-6-trimethylsiloxy-1-TMS-1,2,3,4-tetrahydro-3-pyridine[N,N-bis(TMS)-carboxamide] (*395*) or an isomer of *395* that can directly be converted to yield a variety of piperidine (*397, 398, 400*) and pyridine compounds (*399a, 399b*). *392* and *394* react in the same pattern. Carbon suboxide (*392*) is generated by pyrolysis of bis-(TMS)-malonate (*391*) (Scheme 58).

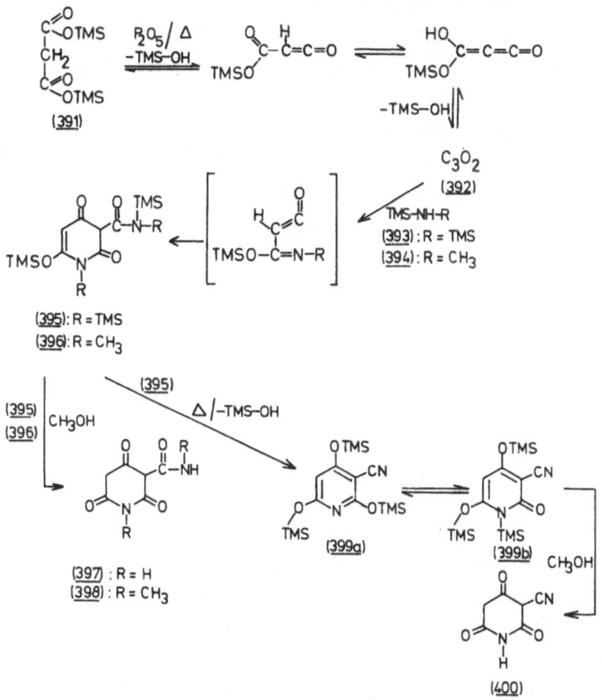

Scheme 58

Thus, *395* gives via pyrolysis 2,4,6-tris(trimethylsiloxy)-3-pyridinecarbonitrile (*399*) and subsequent hydrolysis 2,4,6-trioxo-3-piperidine carbonitrile (*400*) whereas direct hydrolysis of *395* leads to 2,4,6-trioxopiperidinecarboxamide (*397*). Analogously, *396* affords by direct hydrolysis 1-methyl-2,4,6-trioxo-3-piperidine[N-methylcarboxamide] (*398*).

f) Diazines (Pyridazines)

In a different pattern, by using silylated acetylenes, substituted pyridazines are obtainable[217] from the tetrazine derivative *401* in a diene-type reaction, first introduced by Carboni and Lindsey[218]. Via this reaction 4-TMS- (*402*) and 4,5-bis(TMS)-3,6-bis(methoxycarbonyl)pyridazine (*403*) can be achieved in very high yield, being inert against acid catalyzed desilylation (Scheme 59).

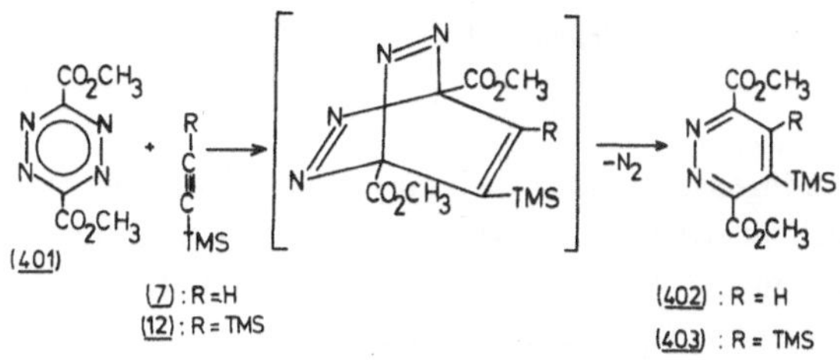

Scheme 59

2 Silylation of Heterocyclic Compounds

Besides the methods already described for synthesizing silylated heterocycles via cyclisation reactions another possible alternative is the direct silylation of heterocyclic substances. Very early investigations[219] showed that the use of silylated amines[219] and silylated carboxylic acid amides[220] is a very appropriate choice[221] when the direct silylation[222, 223] by means of a chlorosilane fails.

Thus, many silylated (especially nitrogen-containing) compounds, have been synthesized[2, 224]. These then undergo further reactions like alkylation[224] (Scheme 60), e.g., tetrakis-TES-uric acid (*404*) gives tetramethyl uric acid (*405*) by means of methyl iodide.

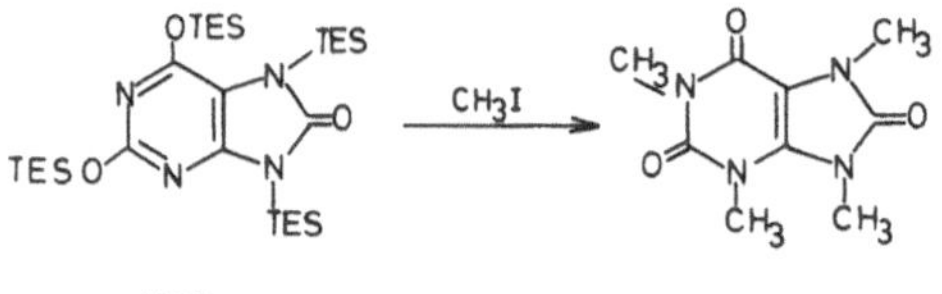

(404) (405) **Scheme 60**

After it was demonstrated that silylated sugars easily form diglycosides[225] when they are applied to 1-halo-sugars, further studies followed. It became apparent that these silylated heterocycles are convenient intermediates, especially for the synthesis of nucleosides when acylated 1-halo-sugar are employed[224, 225].

With 2,3,4,6-tetraacetyl-1-bromo-glucose (*407*), 2-trimethylsiloxypyridine (*406*) and tetrakis-TES-uric acid (*404*) afforded α-pyridone-N-tetraacetylglucoside (*408*) and uric acid-3-glucoside (*409*), respectively, (Scheme 61). This "Hilbert-Johnson"[226] type reaction has many advantages because of its high yields and the fact that it is a one step reaction with excellent reactivity. In the meantime, the number of publications has exorbitantly increased[227]. Parallel to the above described results not only pyranosides but also furanosides have been synthesized, too.

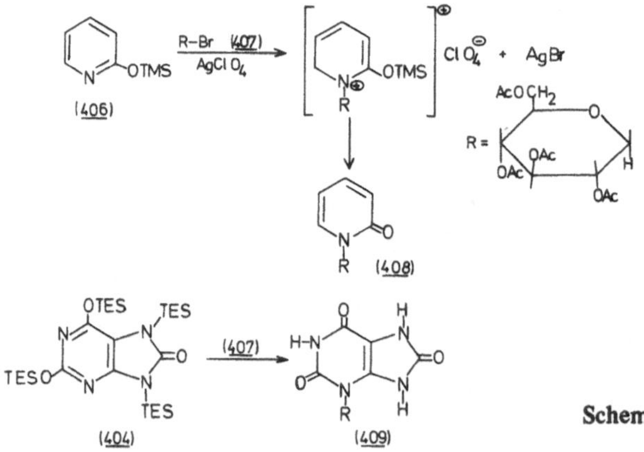

Scheme 61

E Silylated Reagents

1 Trimethylsilyl Azide

Besides the ring-forming reactions already described, trimethylsilyl azide (*14*) gives an access to a great variety of products. As an important property the excellent stability of *14* must be noted[228, 229] which makes it more advantageous to employ TMSA (*14*) instead of other azides.

The reactions of TMSA (*14*) with aldehydes[165, 230] gives α-trimethylsiloxy-alkyl azides (*246, 413–417*); they are highly stable against hydrolysis and higher temperatures (Scheme 62). Thermolysis of the azides *246, 413–417* gives directly N-TMS-butyr- (*246*), valero- (*413*), capro- (*414*)-, isobutyr- (*415*)-, pival- (*416*) and trichloroacetamide (*417*), respectively.

Aldeh	R	Azide	Amide
(240)	$CH_3(CH_2)_2$	(246)	(418)
(410)	$CH_3(CH_2)_3$	(413)	(419)
(411)	$CH_3(CH_2)_4$	(414)	(420)
(234)	$(CH_3)_2CH$	(415)	(421)
(412)	$(CH_3)_3C$	(416)	(422)
(239)	Cl_3C	(417)	(423)

Scheme 62

If dimethyl acetylenedicarboxylate (*350*) is added to *234* and *412* the formation of dimethyl 1-[2-methyl-1-(trimethylsiloxy)propyl]- (*424*) and dimethyl 1-[2,2-dimethyl-1-(trimethylsiloxy)propyl]-1,2,3-triazoledicarboxylate (*425*) is observed; via hydrolysis dimethyl 1,2,3-triazole-4,5-dicarboxylate (*426*) is obtained. *14* and cyclohexene oxide (*427*) gives trans-2-(trimethylsiloxy)cyclohexyl azide (*428*) which undergoes various reactions: hydrolysis, hydrogenation, to form 2-amino-cyclohexanol (*430*) which is converted via diazotization and ring rearrangement to cyclopentane-aldehyde (*431*) – a proof for the trans-conformation of *430* (according to McCasland)[231, 230]. A similar reaction with styrene oxide (*109*) yields 2-phenyl-2-(trimethylsiloxy)ethyl azide (*432*), with *350* dimethyl 1-(2-phenyl-2-trimethylsiloxy-ethyl)-1,2,3-triazole-4,5-dicarboxylate (*436*) is formed which is then hydrolyzed to *426*. Direct hydrolysis of *432* gives 2-phenyl-2-hydroxy-ethyl azide (*433*) which leads to 2-hydroxy-2-phenyl-ethylamine (*434*) via hydrogenation. The hydroxyl group can be silylated to yield 2-phenyl-2-trimethylsiloxyethylamine (*435*)[230] (Scheme 63).

The reaction of *14* with triorganylphosphines gives the (TMS-imino)-triorganyl-phosphoranes[232, 233] *439, 445, 447, 452* while evolving nitrogen.

Hydrolysis of *439, 445–447* initiates desilylation to the corresponding imides *440, 448–450*[232, 233] whereas treatment of *439* with acetyl chloride or acetic anhydride yields acetyliminotriphenylphosphorane (*441*)[234, 235]. The reaction with

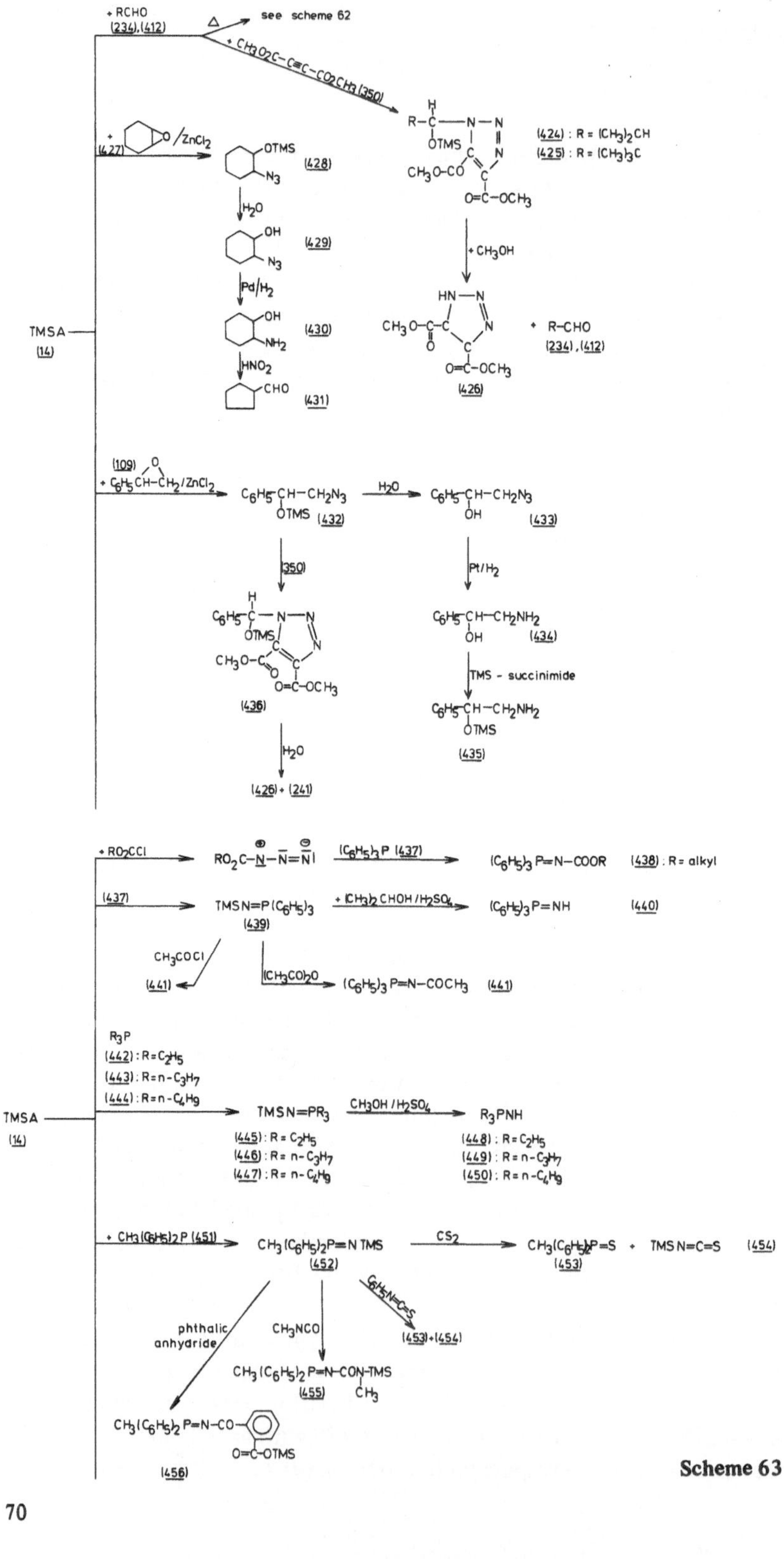

TMSA (14)

+ RCHO (234), (412) → △ → see scheme 62

+ $CH_3O_2C-C{\equiv}C-CO_2CH_3$ (350) → (424): R = $(CH_3)_2CH$; (425): R = $(CH_3)_3C$

+ CH_3OH → (426) + R-CHO (234), (412)

+ cyclohexene oxide / $ZnCl_2$ (427) → (428) → H_2O → (429) → Pd/H_2 → (430) → HNO_2 → (431)

(109) + $C_6H_5CH-CH_2$ (epoxide) / $ZnCl_2$ → $C_6H_5-CH(OTMS)-CH_2N_3$ (432) → H_2O → $C_6H_5-CH(OH)-CH_2N_3$ (433) → Pt/H_2 → $C_6H_5-CH(OH)-CH_2NH_2$ (434) → TMS - succinimide → $C_6H_5-CH(OTMS)-CH_2NH_2$ (435)

(432) → (350) → (436) → H_2O → (426) + (241)

+ RO_2CCl → $RO_2C-N-N=N$ → $(C_6H_5)_3P$ (437) → $(C_6H_5)_3P{=}N-COOR$ (438): R = alkyl

(437) → $TMSN{=}P(C_6H_5)_3$ (439) → + $(CH_3)_2CHOH/H_2SO_4$ → $(C_6H_5)_3P{=}NH$ (440)

(439) → CH_3COCl → (441)

(439) → $(CH_3CO)_2O$ → $(C_6H_5)_3P{=}N-COCH_3$ (441)

R_3P (442): R = C_2H_5; (443): R = n-C_3H_7; (444): R = n-C_4H_9 → $TMSN{=}PR_3$ (445): R = C_2H_5; (446): R = n-C_3H_7; (447): R = n-C_4H_9 → CH_3OH/H_2SO_4 → R_3PNH (448): R = C_2H_5; (449): R = n-C_3H_7; (450): R = n-C_4H_9

+ $CH_3(C_6H_5)_2P$ (451) → $CH_3(C_6H_5)_2P{=}N\,TMS$ (452) → CS_2 → $CH_3(C_6H_5)_2P{=}S$ (453) + $TMSN{=}C{=}S$ (454)

(452) → $C_6H_5N{=}C{=}S$ → (453) + (454)

(452) → CH_3NCO → $CH_3(C_6H_5)_2P{=}N-CON(CH_3)-TMS$ (455)

(452) → phthalic anhydride → $CH_3(C_6H_5)_2P{=}N-CO-C_6H_4-C(=O)-OTMS$ (456)

Scheme 63

452 and phthalic anhydride affords {[ortho-(trimethylsiloxycarbonyl) benzoyl]-imino} -methyldiphenylphosphorane (*456*)[235]. Quite different is the course of reaction when either methyl isocyanate or on the other hand phenyl isothiocyanate is employed. The former leads to [N-methyl-N-TMS-carbamoyl-imido]methyldiphenylphosphorane (*455*)[235] whereas the latter one under drastic conditions (100 °C, 4d) primarily favours the formation of TMS-isothiocyante (*454*)[235]. The same product is obtained if carbon disulfide is employed whereas the by-product in both cases is sulfino-methyl-diphenyl-phosphorane (*453*). Similarly, triphenylphosphinimino-carbaminic acid esters (*438*) can be synthesized by using alkyl chloroformates as reagent[234] (Scheme 64).

Another interesting aspect is the elegant one-step synthesis of manifold substituted isocyanates achieved simultaneously by Washburne et al.[236, 237] and Krichel-dorf[238, 239] who both used acyl chlorides[236–238], acid anhydrides[236] and esters[239] to prepare isocyanates (*457*) via a Curtius rearrangement. This synthesis is especially useful since it enables for instance long chain isocyanates[237] to be obtained (from fatty acid chlorides) and perfluorated compounds[240]. If a perfluoro-

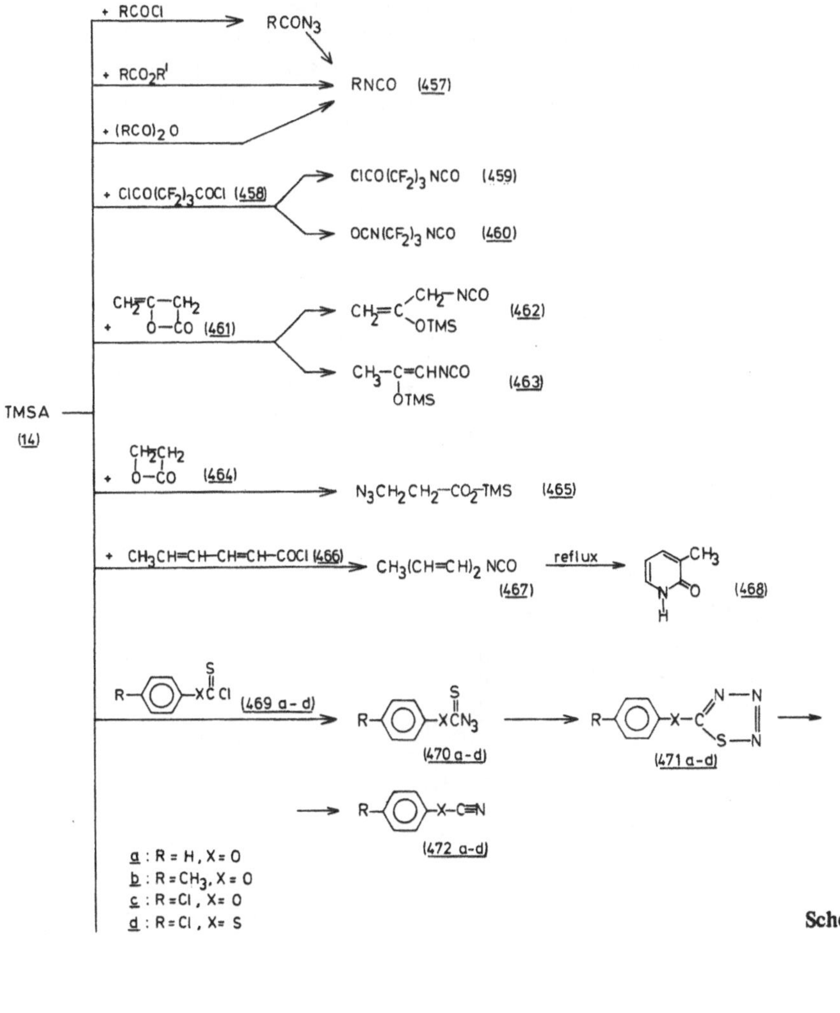

TMSA (14)
+ RCOCl → $RCON_3$ → RNCO (457)
+ RCO_2R' → RNCO (457)
+ $(RCO)_2O$ → RNCO (457)
+ $ClCO(CF_2)_3COCl$ (458) → $ClCO(CF_2)_3NCO$ (459); $OCN(CF_2)_3NCO$ (460)
+ $CH_2{=}C{-}CH_2{-}O{-}CO$ (461) → $CH_2{=}C(CH_2{-}NCO)(OTMS)$ (462); $CH_3{-}C(OTMS){=}CHNCO$ (463)
+ $CH_2CH_2{-}O{-}CO$ (464) → $N_3CH_2CH_2{-}CO_2{-}TMS$ (465)
+ $CH_3CH{=}CH{-}CH{=}CH{-}COCl$ (466) → $CH_3(CH{=}CH)_2NCO$ (467) —reflux→ (468)
+ $R{-}C_6H_4{-}XC(S)Cl$ (469 a-d) → $R{-}C_6H_4{-}XC(S)N_3$ (470 a-d) → $R{-}C_6H_4{-}X{-}C$ (thiatriazole) (471 a-d) → $R{-}C_6H_4{-}X{-}C{\equiv}N$ (472 a-d)

a: R = H, X = O
b: R = CH_3, X = O
c: R = Cl, X = O
d: R = Cl, X = S

Scheme 64

dicarboxylic acid chloride is employed (e.g. perfluoroglutaric chloride (*458*)), the mono- (*459*) and bis-isocyanate (*460*) can be obtained. Two isomeric products occur when diketene (*461*) is substrate, yielding 2-methyl-2-trimethylsiloxyethene isocyanate (*463*) and 2-trimethylsiloxy-allyl-isocyanate (*462*)[239], whereas propiolactone (*464*) gives TMS β-azido-propionate (*465*)[239].

Moreover, sorboyl chloride (*466*) was transformed into the corresponding isocyanate (*467*) and via subsequent cyclization, 3-methyl-2-pyridone (*468*) could be isolated[241].

Cyanates and thiocyanates can be prepared in an analogous way when aryl chlorothioformates (*469a–c*) and aryl chloro-dithio-formates (*469d*) are treated with *14*[242]. Those acyl azides resulting in the first step (*470a–d*) undergo intramolecular cyclization to the corresponding 1-thia-2,3,4-triazoles (*471a–d*) which then under ring cleavage afford the desired cyanates (*472a–c*) and thiocyanate (*472d*)[242] (Scheme 65).

Another type of reaction deserves our attention, the insertion type reactions[243, 244] reported by Washburne and coworkers (Scheme 66).

In the first step by taking *14* and maleic anhydride 3-TMS-1,3-oxazino-2,6(6H)-dione (*473*) is formed which is then hydrolyzed to 1,3-oxazino-2,6(6H)dione (*474*)[243]; equally N-n-butyl-isomaleimide (*475*) gives 3-n-butyl-uracil (*476*)[243].

The situation is different when phthalic anhydride is taken, as primarily an equilibrium of the open (*477a*) and cyclic form (*477b*) is observed that leads under hyd-

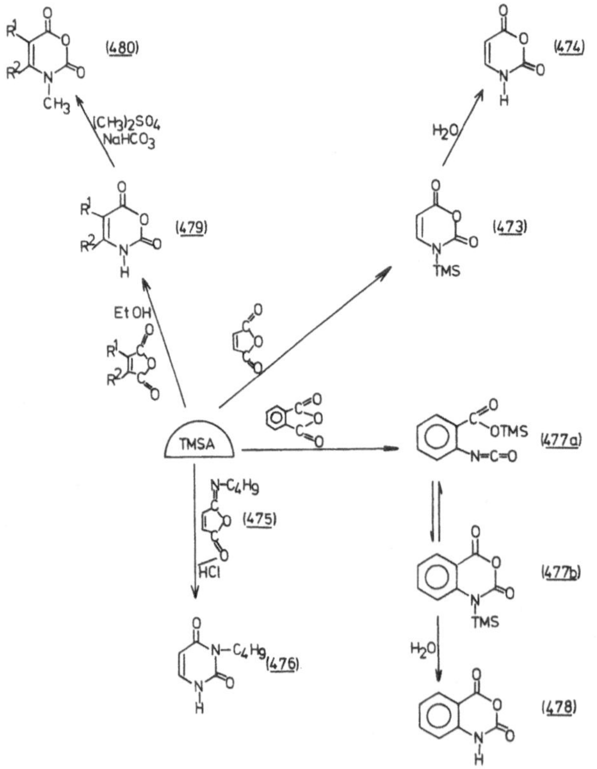

Scheme 65

rolytic desilylation to isatoic anhydride (*478*)[243]. Variously substituted maleic anhydrides afford the corresponding 4,5-disubstituted 1,3-oxazino-2,6-diones (*479*)[244] which can then be methylated by dimethylsulfate to 4,5-disubstituted-3-methyl-1,3-oxazino-2,6-diones (*480*) (Scheme 66).

More contributions of these reactions to the organophosphorus chemistry have been described. *14* and sulfur trioxide form azido-TMS-sulfate (*481*) which, if treated with triphenylphosphine and after subsequent hydrolysis of *482*, gives rise to triphenylphosphoranimino-sulfonic acid (*483*)[245].

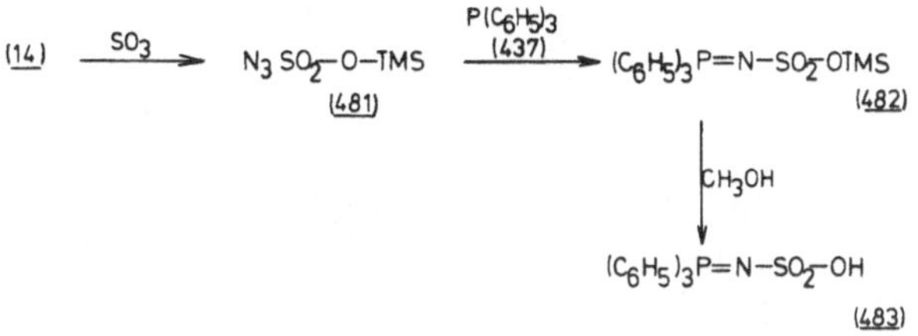

(14) $\xrightarrow{SO_3}$ N_3SO_2–O–TMS (481) $\xrightarrow{P(C_6H_5)_3\ (437)}$ $(C_6H_5)_3P{=}N{-}SO_2{-}OTMS$ (482) $\xrightarrow{CH_3OH}$ $(C_6H_5)_3P{=}N{-}SO_2{-}OH$ (483)

Scheme 66

Another facile reagent combination is the $Pb(OAc)_4$-TMSA system[246] that gives access to α-ketoazides[247, 248], e.g. an α-azido-cyclohexanone (*484*)[247] can be obtained (Scheme 68)[247] and α-azido-acetophenone (*485*) from styrene[248] using the slightly modified $C_6H_5I\text{-}(OAc)_2$-TMSA reagent system.

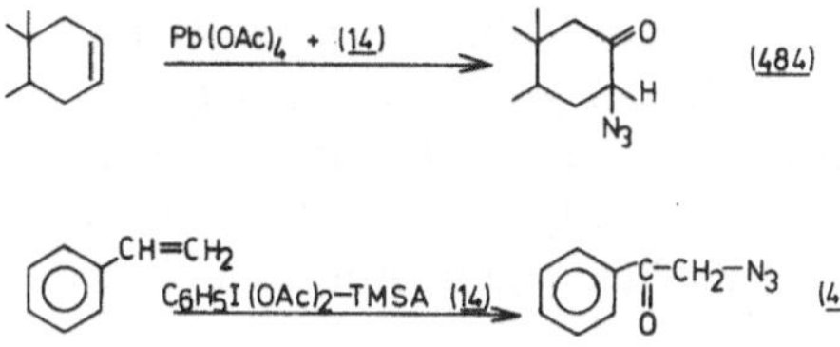

$Pb(OAc)_4$ + (14) → (484)

$C_6H_5CH{=}CH_2$ $\xrightarrow{C_6H_5I(OAc)_2\text{–TMSA}\ (14)}$ $C_6H_5C(O){-}CH_2{-}N_3$ (485)

Scheme 67

Further insertion reactions are worth mentioning. Trans-2-TMS-styrene (*63*) and triethylvinylsilane (*61a*) form with *14* the corresponding bis-silylamines: trans-2-

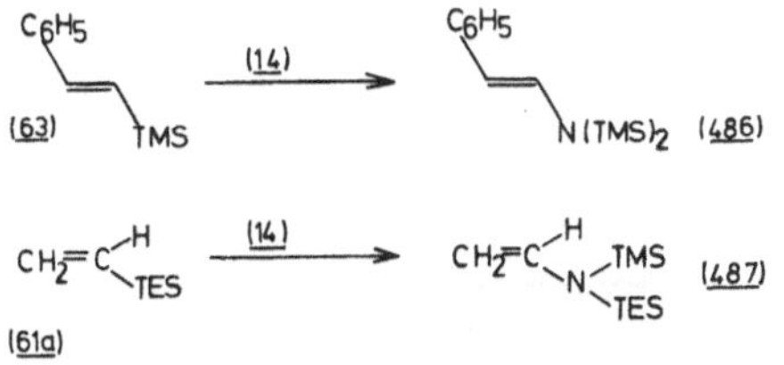

(63) $\xrightarrow{(14)}$ $C_6H_5CH{=}CH{-}N(TMS)_2$ (486)

$CH_2{=}C(H){-}TES$ (61a) $\xrightarrow{(14)}$ $CH_2{=}C(H){-}N(TMS)(TES)$ (487)

Scheme 68

N,N-bis(TMS)-amino styrene (*486*) and N-TES-N-TMS-vinyl-amine (*487*)[249], respectively. TMSA (*14*) produces via UV-photolysis in Ar, N_2 and CO matrices at 10 °K the following rearranged silylimines (*488*) and (*489*)[250]:

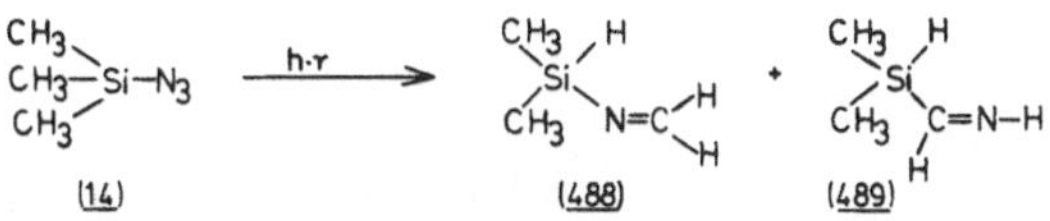

$(CH_3)_3Si{-}N_3$ (14) $\xrightarrow{h\nu}$ $(CH_3)_2Si(H){-}N{=}CH_2$ (488) + $(CH_3)_2Si(H){-}C(H){=}N{-}H$ (489)

Scheme 69

2 Silylation Procedures

a) The Common Silylation

The main classical silylation reactions[2, 9, 10, 251] have not drastically changed in the last decade. A great number of theoretical papers have analyzed the structure of silylating reagents, e.g. in contrast to bis(TMS)-acetamide (*490*) – that has predominantly

$CH_3-C(=O)-N(TMS)-TMS$ (490a) ⇌ $CH_3-C(OTMS)=N-TMS$ (490b) $H-C(=O)-N(TMS)_2$ (491) **Scheme 70**

the N,O structure (*490b*)[220, 221, 252] – bis(TMS)-formamide (*491*) is a N,N-disilylamide[253], as is also the bis(haloalkyl-dimethylsilyl)acetamides[254, 255]. Many papers examining, mechanistic and kinetic problems[256], as well as hindered rotation in silylamides and silylacetanilides[257] have recently been published.

The great breakthrough for silylation methods was its great analytical applicability. It could be shown that silyl derivatives[258] of alcohols, phenols, carboxylic acids (incl. fatty acids), amino acids, amines and carbohydrates are useful tools for analyzing mixtures of several unknown compounds via GLC by means of their retention time[251, 258].

Moreover, it must be noted that silylation has still great synthetic potential because of its easy availability and good yields. Thus, using silyl esters[259] is very advantageous for a peptide synthesis without racemisation. The treatment of Z-amino acids or Z-peptides (as p-nitrophenylesters (*492*)) with N-silylamino acid TMS-esters (*493*) leads to Z-peptide-TMS-esters (*494*) and after subsequent hydrolysis to Z-peptides (*495*) in excellent yields (Scheme 71).

Z.NH–CH(R)–C(=O)–O–C_6H_4–NO_2 (492) + TMS–NH–CH(R')–C(=O)OTMS (493) ⟶

Z.NH–CH(R)–C(=O)–NH–CH(R')–C(=O)OTMS (494) $\xrightarrow[NaHCO_3]{H_2O}$ Z.NH–CH(R)–C(=O)–NH–CH(R')–C(=O)OH (495) **Scheme 71**

An alternative procedure is the direct silylation[259] of Z-amino acids or Z-peptides via TMS-acetamides in presence of catalytic amounts of sulfuric acid (in THF as solvent) at 40–60 °C (Scheme 72) which gives the amino acid TMS-ester (*496*), reacting furthermore with the p-nitrophenylester of a Z-amino-acid or Z-peptide

$H_3N^{\oplus}$–CH(R)–$C(=O)O^{\ominus}$ + TMS–N(R)–C(=O)–R' ⟶ H_2N–CH(R)–C(=O)OTMS (496) + R'–C(=O)–NH–R **Scheme 72**

(*492*) to produce Z-peptide silylester (*494*) whereas the p-nitrophenol generated is immediately transformed into the p-nitrophenyl-TMS-ether by employing a second equivalent of N-TMS-acetamide (Scheme 73).

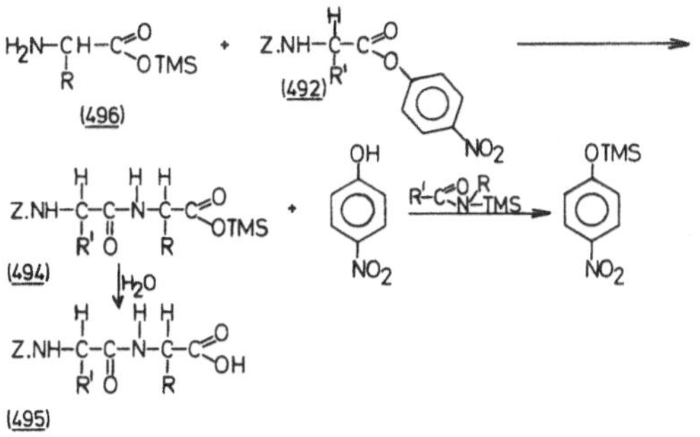

Scheme 73

After hydrolysis the peptide *495* is obtained in good yields without any racemisation – a matter that is highly appreciated[259] – thus, Z-gly-L-Phe-glycine was specifically synthesized.

This method is a fine extension of previous successful attempts[260, 261].

A further application of silylated amino acids is the formation of β-lactams by treating N-TMS-β-amino acid-TMS-esters (*497*) with Grignard reagents under cyclization[262] (Scheme 74). From N-TMS-α-phenyl-β-alanin-TMS-ester (*497*) via the silylated product (*498*), 3-phenyl-2-azetidinone (*499*) can be obtained.

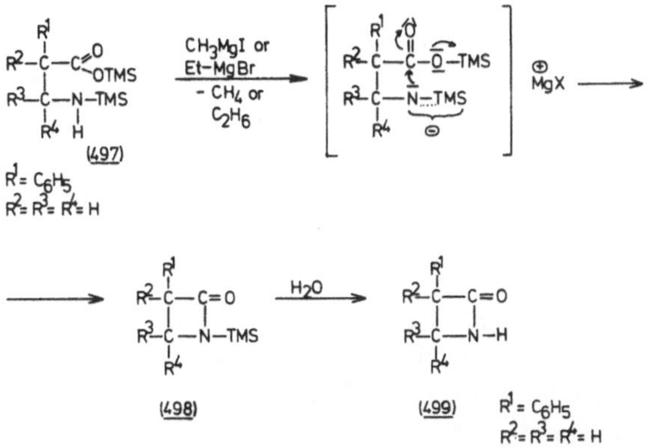

Scheme 74

This reaction is a consequent continuation of an already described lactone synthesis[263] where β- or γ-halo-esters were treated with silver isocyanate to form the corresponding β- or γ-lactone (Scheme 75). Via this method TMS β-bromo-propionate (*500*) affords β-propiolactone (*464*)[263].

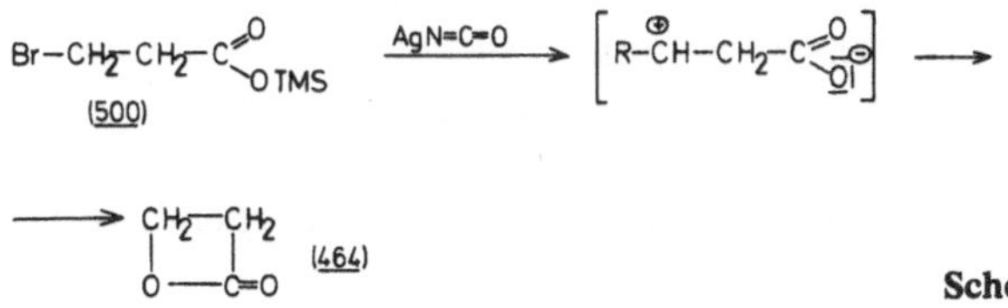

Scheme 75

Besides the already described procedures, another pathway for preparing β-lactams is known[264] using N-TMS-imines. Here, diphenylketene (*501*) and N-TMS-benzaldimines (*502*)[265] yield 3,3,4-triphenyl-2-azetidinone (*503*) (Scheme 76).

$$(C_6H_5)_2C{=}C{=}O\ (501) + C_6H_5(H)C{=}N{-}TMS\ (502) \longrightarrow \text{1-TMS-3,3,4-triphenyl-2-azetidinone} \xrightarrow{H_2O} \text{3,3,4-triphenyl-2-azetidinone}\ (503)$$

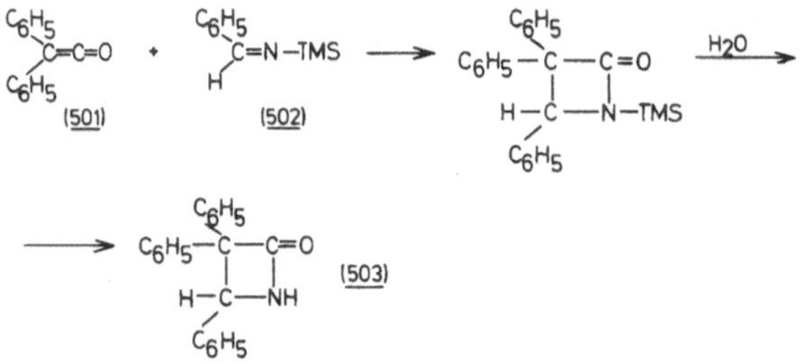

Scheme 76

Concluding this chapter a very convenient silylation of long chain alcohols has been reported[266] employing hexamethyldisiloxane (*504*) in presence of pyridinium-p-toluenesulfonate (*505*) (Scheme 77). The silylated alcohol *506* can be achieved in good yield. The water is removed by means of a Soxhlett extractor, filled with molecular sieves (4 Å).

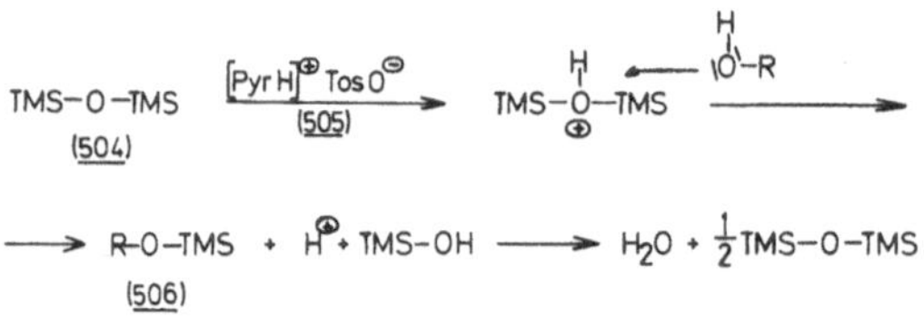

$$TMS{-}O{-}TMS\ (504) \xrightarrow{[PyrH]^{\oplus}\,TosO^{\ominus}\ (505)} TMS{-}\overset{H}{\underset{\oplus}{O}}{-}TMS \xleftarrow{\ H{-}\ddot{O}{-}R\ } \longrightarrow R{-}O{-}TMS\ (506) + H^{\oplus} + TMS{-}OH \longrightarrow H_2O + \tfrac{1}{2}\,TMS{-}O{-}TMS$$

Scheme 77

b) The t-Butyl-dimethyl-silyl-Protection Group

While mainly the trimethylsilyl grouping has found synthetic use for the protection of hydroxyl groups, especially in natural compounds[267–269]. Corey and co-workers have applied the t-butyl-dimethylsilyl group (originally mentioned by Stork and Hudrlik[123]) being highly stable against deblocking via hydrolysis or oxidation – in their recent attempts for a total synthesis of prostaglandines[270, 271] – and later in the terpenoid chemistry by other authors[272]. Because of its bulkiness this moiety

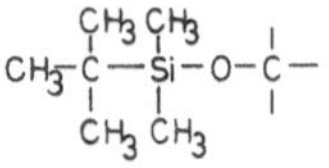

$$(CH_3)_3C{-}Si(CH_3)_2{-}O{-}\overset{|}{\underset{|}{C}}{-} \quad (507)$$

is relatively inert[273] against weak acidic or basic hydrolysis and mild oxidizing and/or reducing conditions (including hydrogenolysis by means of Pd), against $LiAlH_4$[274] and Grignard reagents[272]. The hydrolysis can be done with acetic acid, $(n\text{-}but)_4NF$ (using THF as solvent)[270, 275] or acetic anhydride in presence of $FeCl_3$[276].

Furthermore, this blocking agent has found access to carbohydrate chemistry as protection group[277–282].

A very simple preparation is the reaction of t-butyl-dimethyl-silyl chloride (*508*) with the substrate in presence of imidazole using DMF as solvent[270, 283] (Scheme 78).

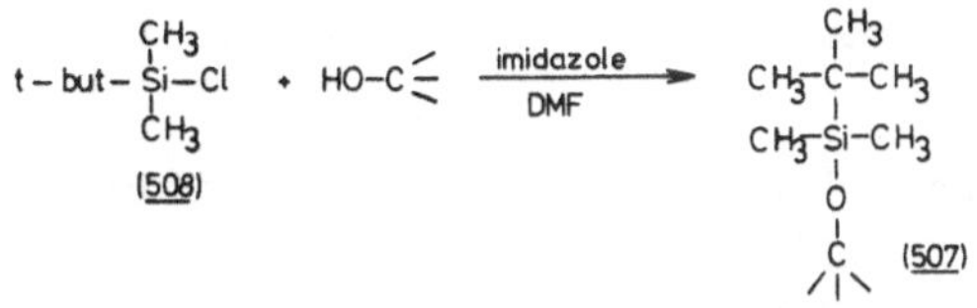

$$\text{t-but-}\underset{CH_3}{\overset{CH_3}{Si}}\text{-Cl} \quad (508) \quad + \quad HO\text{-}C\lessgtr \xrightarrow[DMF]{imidazole} CH_3\text{-}\underset{}{\overset{CH_3}{C}}\text{-}CH_3 \text{ / } CH_3\text{-}Si\text{-}CH_3 \text{ / } O \text{ / } C\lessgtr \quad (507)$$

Scheme 78

The lack of introduction of further chirality makes this protection group a highly valuable tool in the synthetic organic chemistry.

c) N,O-Bis-TMS-carbamate

Early investigations about the reactions of TMS-amines with carbon dioxide and carbon disulfide have led to N,N-disubstituted-TMS-carbamates (*510, 511*) (Scheme 79)[284]. TMS-diethylamine (*509*) gave the two carbamates *510* and *511*, and subsequent treatment with phosphorus pentachloride or thionylchloride yielded N,N-diethyl-carbamoyl- (*512*) and -thiocarbamoylchloride (*513*). However further compounds could not be synthesized by this method.

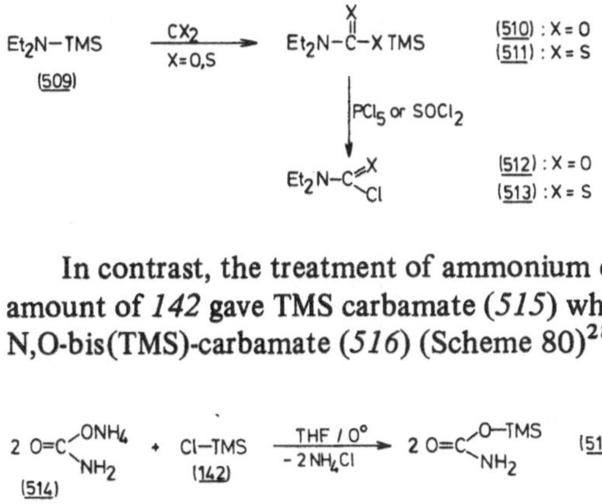

$$Et_2N\text{-}TMS \; (509) \xrightarrow[X=O,S]{CX_2} Et_2N\text{-}\overset{X}{\overset{\|}{C}}\text{-}X\,TMS \quad (510): X=O \quad (511): X=S$$

$$\downarrow PCl_5 \text{ or } SOCl_2$$

$$Et_2N\text{-}C\overset{\nearrow X}{\searrow Cl} \quad (512): X=O \quad (513): X=S$$

Scheme 79

In contrast, the treatment of ammonium carbamate (*514*) with the equivalent amount of *142* gave TMS carbamate (*515*) which decomposed under formation of N,O-bis(TMS)-carbamate (*516*) (Scheme 80)[285]. Intensive studies[286] revealed

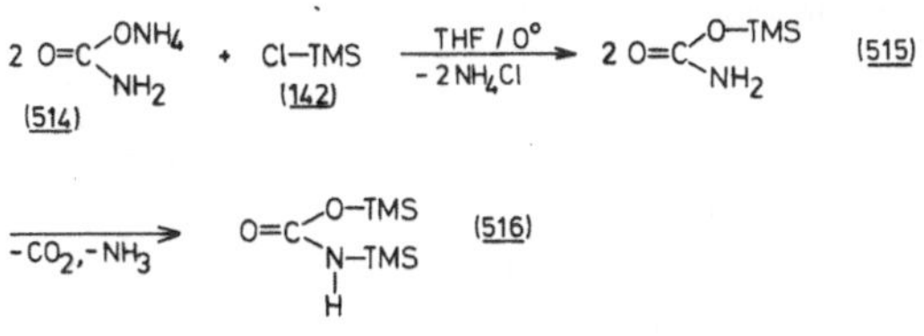

$$2\; O{=}C\overset{ONH_4}{\underset{NH_2}{<}} \; (514) \; + \; Cl\text{-}TMS \; (142) \xrightarrow[-2\,NH_4Cl]{THF\,/\,0^\circ} 2\; O{=}C\overset{O\text{-}TMS}{\underset{NH_2}{<}} \; (515)$$

$$\xrightarrow[-CO_2,\,-NH_3]{} O{=}C\overset{O\text{-}TMS}{\underset{N(H)\text{-}TMS}{<}} \; (516)$$

Scheme 80

that *516* is an extremely suitable silylating agent – protecting alcohols, phenols, carboxylic acids and other substrates[287, 288] (Scheme 81), evolving only carbon dioxide and ammonia. Furthermore, the homologous N,O-bis(TES)-carbamate (*518*)

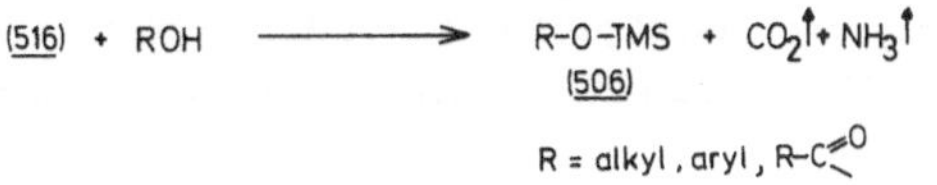

$$(516) + ROH \longrightarrow R\text{-}O\text{-}TMS \; (506) + CO_2\uparrow + NH_3\uparrow$$

$$R = \text{alkyl, aryl, } R\text{-}C\overset{\nearrow O}{\searrow}$$

Scheme 81

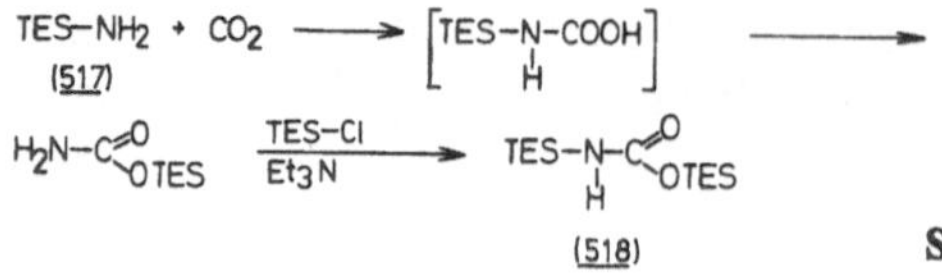

$$TES{-}NH_2 + CO_2 \longrightarrow [TES{-}NH{-}COOH] \longrightarrow$$

(517)

$$H_2N{-}C({=}O)OTES \xrightarrow[Et_3N]{TES{-}Cl} TES{-}NH{-}C({=}O)OTES$$

(518)

Scheme 82

was synthesized, but via a different sequence of reaction (Scheme 82). At first carbon dioxide is pressed into TES-amine (*517*) and the TES-carbamate is obtained immediately (30 min); addition of TES-Cl/triethylamine yields N,O-bis(TES)-carbamate (*518*), a powerful silylating agent[289].

d) Cyclosilylation

A convenient reagent for cyclosilylation turned out to be hexamethylcyclotrisilazane (*519*)[290]. It forms five-, six- and seven-membered rings lacking any kind of polymers[291] and that in preparative order of magnitude. Whereas other cyclizing silyl reagents have only limited applicability – mainly rigid molecules were suitable substrates for this reaction e.g. steroids, salicylic, thiosalicylic or anthranilic acid)[292–295]. Formerly, the following cyclising reagents were used: dichlorodimethylsilane (*167*), diacetoxydimethylsilane or dimethyldimethoxysilane. They need a cautious handling and have only a small field of application. In contrast, it could be shown that HMCTS (*519*) is a universal reagent – favouring the ring formation without any need for bulky moieties at the substrate (Scheme 83). Thusly, 4-sub-

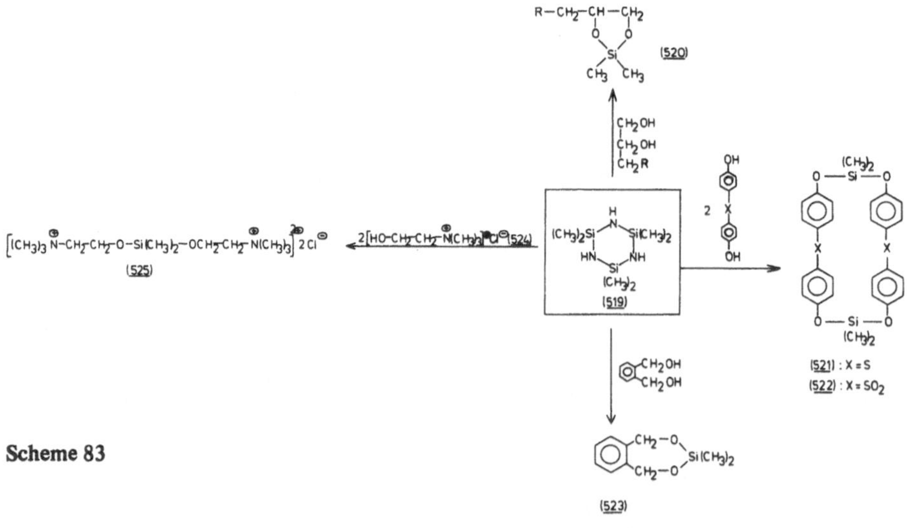

(520) (519) (525) (521) : X = S (522) : X = SO_2 (523)

$2[HO{-}CH_2{-}CH_2{-}\overset{\oplus}{N}(CH_3)_3]Cl^{\ominus}$ (524)

$[(CH_3)_3\overset{\oplus}{N}{-}CH_2{-}CH_2{-}O{-}Si(CH_3)_2{-}OCH_2{-}CH_2{-}\overset{\oplus}{N}(CH_3)_3]\,2Cl^{\ominus}$

Scheme 83

stituted 2,2-dimethyl-1,3-dioxa-2-sila-cyclopentanes (*520*)[298a], [3.1.3.1]paracyclophanes (*521, 522*)[298b], 3,3-dimethyldioxa-1,2,4,5-tetrahydro-3H-3-benzosilepine (*523*)[297] could be obtained. With choline chloride (*524*) (2 equiva.) dimerisation to *525* occurred.

e) Bromo- and Iodo-Trimethylsilane

Although bromo- (*527*) and iodotrimethylsilane (*526*) have been well known for a long time[1, 299, 300]; in synthetic organic chemistry the have proved to be real "sleepers". The situation changed dramatically when in late 1976 the ester and ether cleaving properties of *526* were reported[302, 303, 305] (Scheme 84) which, besides alkyl iodide as a reaction product, generates the TMS-ester and TMS-ether, respectively. Subsequent hydrolysis yields the corresponding carboxylic acid or alcohol (phenol), respectively.

$$R{-}C{\overset{O}{\lessgtr}}_{OR'} \xrightarrow[-R'I]{TMS{-}I\ (526)} R{-}C{\overset{O}{\lessgtr}}_{OTMS} \xrightarrow{H_2O} R{-}C{\overset{O}{\lessgtr}}_{OH} + TMS{-}OH$$

$$R^1{-}O{-}R \xrightarrow{TMS{-}I(526)/H_2O} R{-}I + R^1{-}OH + TMS{-}OH$$

Scheme 84

In the meantime, this reaction was modified for the preparative conversion of alcohols into the corresponding iodides[304].

In contrast to these results, bromo-trimethylsilane (*527*) has only ester cleaving properties referring to dialkyl phosphonates (Scheme 85). Via this reagent a great variety of substituted phosphonic acids can be prepared in hitherto unknown yields.

$$R^1R^2C(H){-}\overset{O}{\overset{\|}{P}}(OR^3)_2 \; (R^3 = \text{alkyl}) \xrightarrow{TMS{-}Br\ (527)} R^1R^2CH{-}\overset{O}{\overset{\|}{P}}(OTMS) + 2R^3{-}Br \; (R^3 = \text{alkyl})$$

$$\xrightarrow{H_2O} R^1R^2CH{-}\overset{O}{\overset{\|}{P}}(OH)_2 + TMS{-}O{-}TMS$$

Scheme 85

f) Cyanosilylation

As already mentioned, the addition of TMS-CN (*255*) across the carbonyl group[171] is a very reasonable reaction (see C, 3). It is not a silylation in the "usual sense" of the term but its versatility and its properties as a protection group gives evidence of many parallels which only silylation normally possesses. Early attempts to synthesize *255* were undertaken by means of a KCN-18-crown-6-complex[308, 309]. Recently, two very simple preparations of *255* have been reported, simplifying the access to *255*[310, 311], and furthermore a "one-pot-synthesis" for the cyanosilylation reaction as a whole[312]. Aldehydes[315], ketones (incl. some α,β-unsaturated ones) react with *255* (in presence of catalytic amounts of ZnI_2)[312, 313, 316]. So do p-quinones[308, 313], dialdehydes, diketones[314] and acetylacetone if an excess of *255* is employed[314]. Cyanide ions have a catalyzing influence[171]. Aldehydes and ketones give cyanohydrins[313, 316], equally epoxides the corresponding β-siloxy-nitriles[317], acid chlorides siloxymalonitriles[316, 317] (vide infra) and chloroformates cyanoformates[317], respectively. Treating quinones with *255* is a fine route for protection and on the other hand gives access to quinols by means of organometallic reagents.

The employment of TMS-CN principally shows the protecting properties of *255*[308, 313, 318]: ketones primarily yield the corresponding trimethylsiloxy-cyano compounds (*528*)[313, 316], methyl chloroformate the methyl cyanoformate (*529*)[317], acetylchloride forms with 2 equivalents *255* methyl-trimethylsiloxy-malodinitrile (*530*)[317], from oxalylchloride and 4 equivalents *255* 1,1,2,2-tetracyano-1,2-bis-(trimethylsiloxy)ethane (*531*) is obtained. 1,2-dimethyloxirane (*532*) gives under ring cleavage 1,2-dimethyl-1-cyano-2-trimethylsiloxy-ethane (*533*). Benzophenone (*135*) in the first step affords diphenyl-cyano-trimethylsiloxy-methane (*538*) which is then treated with lithiumalanate to yield 1,1-diphenyl-1-ethanolamine (*539*)[313, 319] – a course of reaction based on results from Parham and Roosevelt[320]. The formation of carbanions (e.g. *536*) and subsequent alkylation (e.g. *537*) is possible, too.[315, 322]. 2-Trimethylsiloxy-2-pentene-4-one (*190a*) forms two isomers, the 1-(*540*) and 2-alkene (*541*), 2-(trimethylsiloxy)-4-cyano-1-(*540*) and -2-pentene (*541*)[321]. Quinone (*263*) uses *255* as protection group and subsequent alkylation yield quinol *535*[318, 323].

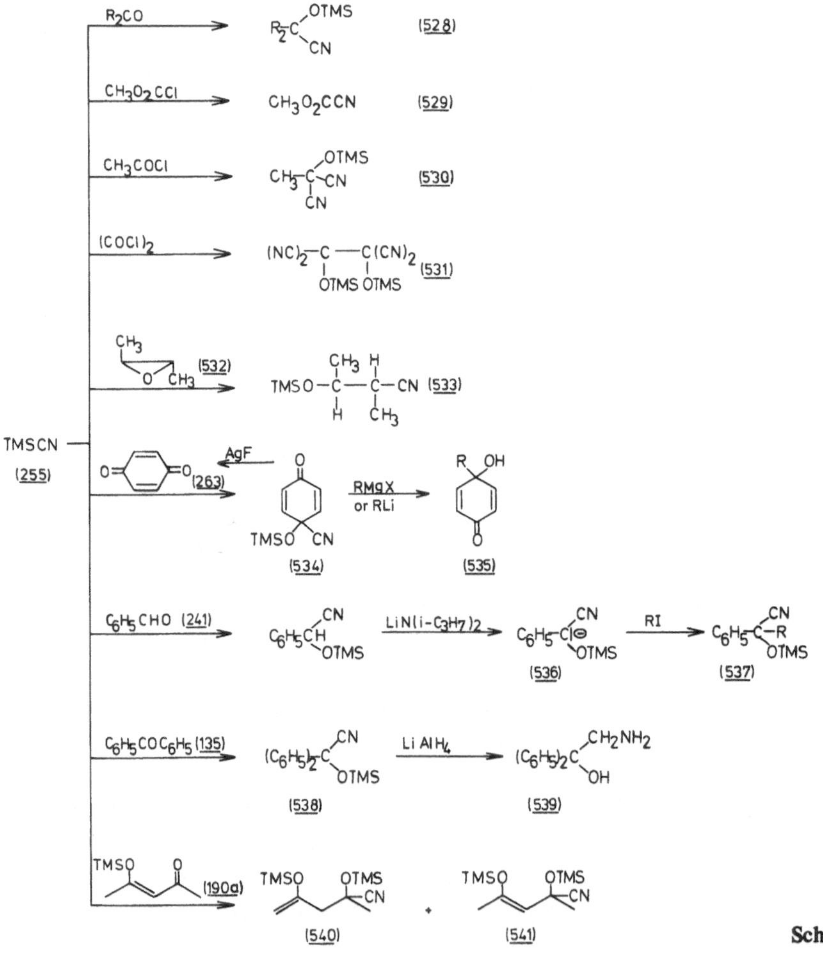

TMSCN (255): R_2CO → $R_2C(OTMS)CN$ (528); CH_3O_2CCl → CH_3O_2CCN (529); CH_3COCl → $CH_3C(OTMS)(CN)_2$ (530); $(COCl)_2$ → $(NC)_2C(OTMS)–C(OTMS)(CN)_2$ (531); 1,2-dimethyloxirane (532) → TMSO–CH(CH_3)–CH(CH_3)–CN (533); quinone (263) → 534 (TMSO, CN) ← AgF; 534 —RMgX or RLi→ 535 (R, OH); C_6H_5CHO (241) → $C_6H_5CH(CN)OTMS$ —LiN(i-C_3H_7)$_2$→ $C_6H_5C^{\ominus}(CN)OTMS$ (536) —RI→ $C_6H_5C(CN)(R)OTMS$ (537); $C_6H_5COC_6H_5$ (135) → $(C_6H_5)_2C(CN)OTMS$ (538) —LiAlH_4→ $(C_6H_5)_2C(CH_2NH_2)OH$ (539); 190a → 540 + 541

Scheme 86

Employment of silver fluoride to *534* has a cleaving effect forming back *263* (Scheme 86).

Just recently, a cyclo-cyanosilylation has been described[321]. Ryu and his co-workers have taken dicyanodimethylsilane (*543*)[310] as a reagent under the conditions of the common cyanosilylation and obtained cyclic silyl enol ether (*544, 545*) when β-hydroxyketones or β-diketones were employed (Scheme 87).

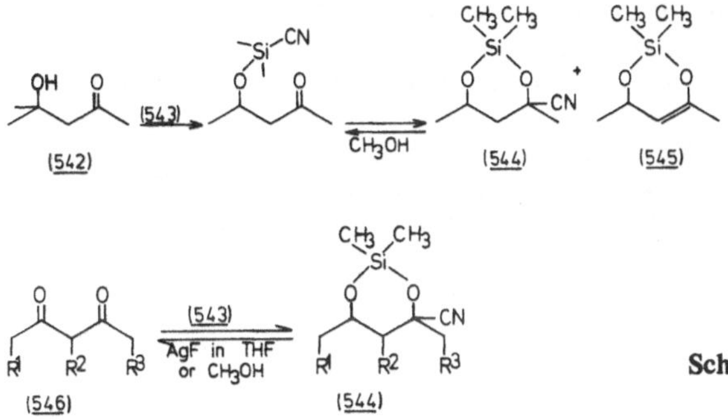

Scheme 87

A β-hydroxyketone (*542*) forms with *543* the corresponding 1,3-dioxa-2-sila-4-cyanocyclohexane (*544*) and the unsaturated isomer *545*. With methanol, de-cyanosilylation is obtained. The β-diketone *546* forms *544*, too; in that case it is better cleaved by AgF/THF or only methanol.

F References

1. Eaborn, C.: "Organosilicon Compounds", Butterworths, London 1960/Academic Press, New York, N. Y. 1960
2. Birkofer, L. and Ritter, A., in: Foerst, W. (Ed.) "Neuere Methoden der präparativen organischen Chemie", Vol. V, p. 185, Verlag Chemie, Weinheim 1967/"Newer Methods in Preparative Organic Chemistry", Vol. V, p. 211, Academic Press, New York, N. Y. 1968
3. Sommer, L. H.: "Stereochemistry, Mechanism and Silicon", McGraw-Hill, New York, N. Y. 1965
4. Petrov, A. D., Mironov, B. F., Ponomarenko, V. A., and Chernyshev, E. A.: "Synthesis of Organosilicon Monomers", Heywood, London 1964/Consultant Bureau, New York, N. Y. 1964
5. Rühlmann, K.: Z. Chem. *5*, 130 (1965)
6. Bažant, V., Chvalovský, V., Rathouský, J., (Ed.): "Organosilicon Compounds", Academic Press, New York, 1965
7. Bažant, V., Chvalovský, V., Rathouský, J.: "Handbook of Organosilicon Compounds", Marcel Dekker, New York, N. Y. 1976
8. Noll, W. (Ed.): "Chemistry and Technology of Silicones", 2nd Edit., Verlag Chemie, Weinheim 1967/Academic Press, New York. N. Y. 1968
9. Pierce, A. E.: "Silylation of Organic Compounds", Pierce Chemical Co, Rockford, Ill., 1968
10. Klebe, J. F.: Acc. Chem. Res. *3*, 299 (1970)
11. Washburne, S. S.: J. Organomet. Chem. *83*, 155 (1974)
12. Fleming, I.: Chem. Ind. (London), *1975*, 449
13. Washburne, S. S.: J. Organomet. Chem. *123*, 1 (1976)

14. Seyferth, D. (Ed.): "New Applications of Organometallic Reagents in Organic Synthesis", J. Organomet. Chem. Library, Vol. 1 and 2., Elsevier, Amsterdam 1976
   a) Hudrlik, P. F.: ibid, Vol. 1. p. 127
   b) Calas, R. and Dunoguès, J.: ibid, Vol. 2, p. 277
15. Colvin, E. W.: Chem. Soc. Rev. *7*, 15 (1978)
16. for IUPAC-D6-Organosilicon-Nomenclature see: Chvalovský, V. and Bláha, K.: Chemické Listy *72*, 618 (1978)
17. Kipping, F. S.: Proc. Chem. Soc. *20*, 15 (1904)
18. Kipping, F. S. and Lloyd, L.: J. Chem. Soc. *91*, 209 (1907)
19. Dilthey, W. and Edouardoff, F.: Ber. dtsch. chem. Ges. *37*, 1139 (1904)
20. Benkeser, R. A. and Hickner, R. A.: J. Amer. Chem. Soc. *80*, 5298 (1958)
21. Benkeser, R. A., Burrous, M. L., Nelson, L. E., Swisher, J. V.: J. Amer. Chem. Soc. *83*, 4385 (1961)
22. Benkeser, R. A.: J. Org. Chem. *32*, 2634 (1967)
23. Tamao, K., Miyako, K., Kiso, Y., Kumada, M.: J. Amer. Chem. Soc. *97*, 5603 (1975)
24. Seyferth, D. (Ed.): "Hydrosilylation", J. Organomet. Chem. Library, Vol. 5, Elsevier, Amsterdam, 1977
25. Hayashi, T., Yamamoto, K., Kumada, M.: Tetrahedron Lett. *1975*, 3
26. Kagan, H. B.: Pure Appl. Chem. *43*, 401 (1975)
27. Ojima, J., Kogure, T., Kumagai, M.: J. Org. Chem. *42*, 1671 (1977)
28. Birkofer, L. and Siegert, K.: Chem. Ber., in press
29. Lehmann, J., Schäfer, H.: Chem. Ber. *105*, 969 (1972)
30. Birkofer, L. and Kühn, Th.: Chem. Ber. *111*, 3119 (1978)
31. Volnov, Y. N., Reutt, A.: Zh. Obshch. Khim. *10*, 1600 (1940) C. A. *35*, 2853 (1941)
32. Petrov, A. D., Schtschukowskaja, L. L.: Dokl. Akad. Nauk. SSSR *86*, 551 (1952)
33. Birkofer, L., Ritter, A., Uhlenbrauck, H.: Chem. Ber. *96*, 3280 (1963)
34. Birkofer, L., Eichstädt, D.: J. Organomet. Chem. *145*, C 29 (1978)
35. Walton, D. R. M., Waugh, F.: J. Organomet. Chem. *37*, 41 (1972)
36. Zweifel, G., Backlund, S. S.: J. Amer. Chem. Soc. *99*, 3184 (1977)
37. Eisch, J. J., Damasevitz, G. A.: J. Org. Chem. *41*, 2214 (1976)
38. Uchida, K., Utimoto, K., Nozaki, H.: J. Org. Chem. *41*, 2215 (1976)
39. Köster, R., Hagelee, L. A.: Synthesis *1976*, 118
40. Uchida, K., Utimoto, K., Nozaki, H.: J. Org. Chem. *41*, 2941 (1976)
41. Obayashi, M., Utimoto, K., Nozaki, H.: Tetrahedron Lett. *1977*, 1805
42. Wertmijze, H., Meijer, J., Vermeer, P.: Tetrahedron Lett., *1977*, 1823
43. Funk, R. L., Vollhardt, K. P. C.: J. C. S. Chem. Commun. *1976*, 833
44. Eastmond, R., Walton, D. R. M.: Tetrahedron *28*, 4591 (1972)
45. Eastmond, R., Johnson, T. R., Walton, D. R. M.: Tetrahedron *28*, 4601 (1972)
46. Harris, S. J., Walton, D. R. M.: Tetrahedron *34*, 1037 (1978)
47. Bhattacharya, S. N., Josiah, B. M., Walton, D. R. M.: Organomet. Chem. Synth. *1*, 145 (1971)
48. Walton, D. R. M., Waugh, F.: J. Organomet. Chem. *37*, 45 (1972)
49. Ghose, B. N., Walton, D. R. M.: Synthesis *1974*, 890
50. Eaborn, C., Walton, D. R. M.: J. Organomet. Chem. *4*, 217 (1966)
51. Newman, H.: J. Org. Chem. *38*, 2254 (1973)
52. Bourgeois, P., Merault, G., Calas, R.: J. Organomet. Chem. *59*, C 4 (1973)
53. Casara, P., Metcalf, B. W.: Tetrahedron Lett. *1978*, 1581
54. Metcalf, B. W., Casara, P.: ibid. *1975*, 3337
55. Metcalf, B. W., Jund, K.: ibid. *1977*, 3689
56. Pillot, J. P., Dunoguès, J., Calas, R.: Compt. Rend. Acad. Sci. *278*, 789 (1974)
57. Fleming, I., Pearce, A.: J. C. S. Chem. Comm. *1975*, 633
58. Pillot, J. P., Dunoguès, J., Calas, R.: Bull. Soc. Chim. France *1975*, 2143
59. Mironov, V. F., Glukhovtsev, V. G., Petrov, A. D.: Dokl. Akad. Nauk. SSSR *104*, 865 (1955)
60. Petrov, A. D., Mironov, V. F., Glukhovtsev, V. G.: Izv. Akad. Nauk SSSR, Otd. Khim. Nauk *1956*, 461

61. Bock, H., Seidl, H.: J. Organomet. Chem. *13,* 87 (1968)
62. Yamamoto, K., Yoshitake, J.: Chem. Lett. *1978,* 859
63. Chan, T. H., Mychaijlowskij, W., Ong, B. S., Harpp, D. N.: J. Organomet. Chem. *107,* C1 (1976)
64. Mychaijlowskij, W., Chan, T. H.: Tetrahedron Lett. *1976,* 4439
65. Utimoto, K., Kitai, M., Nozaki, H.: ibid. *1975,* 2825
66. Miller, R. B., Reichenbach, T.: ibid. *1974,* 543
67. Fritz, G., Grobe, J.: Z. Anorg. Allgem. Chem. *309,* 98 (1961)
68. Sommer, L. H. et al.: J. Amer. Chem. Soc. *76,* 1613 (1954)
69. Eisch, J. J., Foxton, M. W.: J. Org. Chem. *36,* 3520 (1971)
70. Calas, R., Bourgeois, P., Duffaut, N.: Compt. Rend. Acad. Sci. *263,* 243 (1966)
71. Koenig, K. E., Weber, W. P.: Tetrahedron Lett. *1973,* 2533
72. Stork, G., Ganem, B.: J. Amer. Chem. Soc. *95,* 6152 (1973)
73. for a review see: Gawley, R. E.: Synthesis *1976,* 777
74. Stork, G., Colvin, E.: J. Amer. Chem. Soc. *93,* 2080 (1971)
75. Hudrlik, P. F., Peterson, D., Rona, R. J.: J. Org. Chem. *40,* 2263 (1975)
76. Hudrlik, P. F., Misra, R. N., Withers, G. P., Hudrlik, A. M., Rona, R. J., Arcoleo, J. P.: Tetrahedron Lett. *1976,* 1453
77. Hudrlik, P. F., Wan, C.-N., Withers, G. P.: ibid. *1976,* 1449
78. Hudrlik, P. F., Arcoleo, J. P., Schwartz, R. H., Misra, R. N., Rona, R. J.: ibid. *1977,* 591
79. Chan, T. H., Lau, P. W. K., Li, M. P.: ibid. *1976,* 2667
80. Chan, T. H., Lau, P. W. K., Li, M. P.: ibid. *1974,* 3511
81. Bockman jr., R. K.: ibid. *1974,* 3365
82. Robbins, C. M., Whitham, G. H.: J. C. S. Chem. Comm. *1976,* 697
83. Birkofer, L., Bierwirth, E., Ritter, A.: Chem. Ber. *94,* 821 (1961)
84. West, C. T., Donnelly, S. J., Kooistra, D. A., Doyle, M. P.: J. Org. Chem. *38,* 2675 (1973) (review)
85. Kursanov, D. N., et al.: Synthesis *1973,* 421
86. Parnes, Z. N., et al.: Izv. Akad. Nauk SSSR, Ser. Khim. *1973,* 1918
87. Kursanov, D. N., Parnes, Z. N., Loim, N. M.: Synthesis *1974,* 633 (review)
88. Parnes, Z. N., et al.: Zh. Org. Khim. *9,* 1704 (1973)
89. Doyle, M. P., McOsker, C. C., West, C. T.: J. Org. Chem. *41,* 1393 (1976)
90. Adlington, M. G., Orfanopoulos, M., Fry, J. L.: Tetrahedron Lett. *1976,* 2955
91. Doyle, M. P., De Bruyn, D. J., Donnelly, S. J., Kooistra, D. A., Odubela, A. A., West, C. T., Sonnebelt, S. M.: J. Org. Chem. *39,* 2740 (1974)
92. Citron, J. D.: J. Org. Chem. *34,* 1977 (1969)
93. Segall, Y., Granoth, J., Kalir, A.: J. C. S., Chem. Comm. *1974,* 501
94. see also: Marsi, K. L.: J. Org. Chem. *39,* 265 (1974)
95. Serebryakova, T. A., et al.: Izv. Akad. Nauk. SSSR, Ser. Khim. *1973,* 1916
96. Serebryakova, T. A., et al.: Izv. Akad. Nauk. SSSR, Ser. Khim. *1973,* 1917
97. Lipowitz, J., Bowman, S. A.: J. Org. Chem. *38,* 162 (1973)
98. Ojima, J., Kogure, T., Nagai, Y.: Tetrahedron Lett. *1972,* 5035
99. Ojima, J., Kogure, T., Nagai, Y.: ibid. *1973,* 2475
100. Langlois, N., Dang, T. P., Kagan, H. B.: ibid. *1973,* 4865
101. Birkofer, L., Ramadan, N.: Chem. Ber. *104,* 138 (1971)
102. Weyenberg, D. R., Toporcer, L. H.: J. Org. Chem. *30,* 943, (1965)
103. Birkofer, L., Ramadan, N.: J. Organomet. Chem. *44,* C 41 (1972)
104. Birkofer, L., Ramadan, N.: J. Organomet. Chem. *92,* C 41 (1975)
105. Birkofer, L., Ramadan, N.: Chem. Ber. *108,* 3105 (1975)
106. Bolourtchian, M., Saednya, A.: Bull. Soc. Chim. Fr. II *1978,* 170
107. Dunoguès, J., Ekouya, A., Calas, R., Duffaut, N.: J. Organomet. Chem. *87,* 151 (1975)
108. Calas, R., Dunoguès, J. Ekouya, A., Merault, G., Duffaut, N.: J. Organomet. Chem. *65,* C 4 (1975)
109. Laguerre, M., Dunoguès, J., Calas, R.: Tetrahedron Lett. *34,* 1823 (1978)
110. Ekouya, A., Dunoguès, J., Calas, R.: J. Chem. Res. *(S) 1978,* 296

111. Ando, W., Ikeno, M., Sekiguchi, A.: J. Amer. Chem. Soc. *100*, 3613 (1978) see also for a similar compound
112. Dunoguès, J., Calas, R., Ardoin, N.: J. Organomet. Chem. *43*, 127 (1972)
113. Picard, J. P., Dunoguès, J., Calas, R.: J. Organomet. Chem. *77*, 167 (1974)
114. Rasmussen, J. K.: Synthesis *1977*, 91 (review)
115. Petrov, A. D., et al.: Izv. Akad. Nauk. SSSR *1958*, 954
116. Petrov, A. D., Sadykh-Zade, S. I.: Bull. Soc. Chim. France *1959*, 1932
117. Petrov, A. D., Sadykh-Zade, S. I.: Dokl. Akad. Nauk. SSSR *121*, 119 (1958)
118. Sadykh-Zade, S. I., Petrov, A. D.: Zh. Obshch, Khim. *29*, 3194 (1959); J. Gen. Chem. USSR (Engl. Transl.) *29*, 3159 (1959)
119. Duffaut, N., Calas, R.: Compt. Rend. Acad. Sci. *245*, 906 (1957)
120. e.g. see review: Baukov, Yu. J., Lutsenko, L. F.: Organomet. Chem. Rev. *A 6*, 355 (1970)
121. Stork, G., Hudrlik, P. F.: J. Amer. Chem. Soc. *90*, 4462 (1968)
122. Stork, G., Hudrlik, P. F.: J. Amer. Chem. Soc. *90*, 4464 (1968)
123. House, H. O., Czuba, L. J., Gall, M., Olmsted, H. D.: J. Org. Chem. *34*, 2324 (1969)
124. House, H. O., Gall, M., Olmsted, H. D.: J. Org. Chem. *36*, 2361 (1971)
125. Brown, C. A.: J. Org. Chem. *39*, 1324 (1974)
126. Brown, C. A.: ibid. *39*, 3913 (1974)
127. Birkofer, L., Dickopp, H.: Chem. Ber. *102*, 14 (1969)
128. Simchen, G., Kober, W.: Synthesis *1976*, 259
129. see also: Emde, H., Simchen, G.: Synthesis *1977*, 636 (applications to nitriles)
130. Nakamura, E., Murofushi, T., Shimizu, M., Kuwajima, J.: J. Amer. Chem. Soc. *98*, 2346 (1976)
131. Helberg, L. H., Juarez, A.: Tetrahedron Lett. *1974*, 3553
132. Ojima, J., Nagai, Y.: J. Organomet. Chem. *57*, C 42 (1973)
133. Coates, R. M., Landefeer, L. O., Smillie, R. D.: J. Amer. Chem. Soc. *97*, 1619 (1975)
134. Ito, Y., Konoike, T., Saegusa, T.: J. Amer. Chem. Soc. *97*, 649 (1975)
135. Kuroki, Y., Murai, S., Sonoda, N., Tsutsumi, S.: Organomet. Chem. Synth. *1*, 465 (1972)
136. Reuss, R. H., Hassner, A.: J. Org. Chem. *39*, 1785 (1974)
137. Klein, J., Levene, R., Dunkelblum, E.: Tetrahedron Lett. *1972*, 2845
138. Larson, G. L., Hernandez, D., Hernandez, A.: J. Organomet. Chem. *76*, 9 (1974)
139. Larson, G. L., Hernandez, E., Alonso, C., Nieves, J.: Tetrahedron Lett. *1975*, 4005
140. Mukaiyama, T., Banno, K., Narasaka, K.: J. Amer. Chem. Soc. *96*, 7503 (1974)
141. Narasaka, K., Soai, K., Aikawa, Y., Mukaiyama, T.: Bull. Chem. Soc. Japan *49*, 779 (1976)
142. Brook, A. G., McCrae, D. A.: J. Organomet. Chem. *77*, C 19 (1974)
143. Birkofer, L., Ritter, A., Vernaleken, H.: Chem. Ber. *99*, 2518 (1966)
144. Denis, J. M., Girard, C., Conia, J. M.: Synthesis *1972*, 549
145. Simmons, H. E., Smith, R. D.: J. Amer. Chem. Soc. *80*, 5223 (1958)
146. Simmons, H. E., Smith, R. D.: J. Amer. Chem. Soc. *81*, 4256 (1959)
147. cf.: Rawson, R. J., Harrison, J. T.: J. Org. Chem. *35*, 2057 (1970)
148. review: Conia, J. M.: Pure Appl. Chem. *43*, 317 (1975)
149. review: Girard, C., Conia, J. M.: J. Chem. Res. (S) *1978*, 182; J. Chem. Res. (M) *1978*, 2351
150. Denis, J. M., Conia, J. M.: Tetrahedron Lett. *1972*, 4593
151. Murai, S., Aya, T., Sonoda, N.: J. Org. Chem. *38*, 4354 (1973)
152. Rubottom, G. M., Lopez, M. J.: J. Org. Chem. *38*, 2097 (1973)
153. Murai, S., Aya, T., Renge, T., Ryu, I., Sonoda, N.: J. Org. Chem. *39*, 858 (1974)
154. Murai, S., Seki, Y., Sonoda, N.: J. C. S. Chem. Comm. *1974*, 1032
155. Conia, J. M., Girard, C.: Tetrahedron Lett. *1973*, 2767
156. Conia, J. M., Girard, C.: ibid. *1974*, 3327
157. Ryu, I., Murai, S., Otani, S., Sonoda, N.: ibid. *1977*, 1995
158. Le Goaller, R., Pierre, J.-L.: Bull. Soc. Chim. Fr. *1973*, 1531
159. Girard, C., Amice, P., Barnier, J. P., Conia, J. M.: Tetrahedron Lett. *1974*, 3329
160. Barnier, J. P., Garnier, B., Girard, C., Denis, J. M., Salaun, J. R., Conia, J. M.: ibid. *1973*, 1747

161. review: Conia, J. M., Robson, M. J.: Angew. Chem. *87,* 505 (1975); Int. Ed. *14,* 473 (1975)
162. Girard, C., Conia, J. M.: Tetrahedron Lett. *1974,* 3333
163. Trost, B. M., Kurozumi, S.: ibid. *1974,* 1929
164. Itoh, K., Fukui, M., Ishii, Y.: ibid. *1968,* 3867
165. Birkofer, L., Müller, F., Kaiser, W.: ibid. *1967,* 2781
166. Birkofer, L., Ritter, A., Wieden, H.: Chem. Ber. *95,* 971 (1962)
167. Pinkerton, F. H., Thomas, S. F.: J. Heterocycl. Chem. *6,* 433 (1969)
168. Pinkerton, F. H., Thomas, S. F.: J. Organomet. Chem. *24,* 623 (1970)
169. Webb, A. F., Sethi, D. L., Gilman, H.: J. Organomet. Chem. *21,* P 61 (1970)
170. Ishikawa, N., Isobe, K.: Chem. Lett. *1972,* 435
171. Evans, D. A., Truesdale, L. K.: Tetrahedron Lett. *1973,* 4929
172. Miller, L. L., Stewart, R. F.: J. Org. Chem. *43,* 3078 (1978)
173. Satgé, J., Couret, C., Escudié, J.: J. Organomet. Chem. *30,* C 70 (1971)
174. Couret, C., Satgé, J., Couret, F.: J. Organomet. Chem. *47,* 67 (1973)
175. Couret, C., Escudié, J., Couret, F.: J. Organomet. Chem. *57,* 287 (1973)
176. Couret, C., Escudié, J., Anh, N. T., Soussan, G.: J. Organomet. Chem. *91,* 11 (1975)
177. Evans, D. A., Hurst, K. M., Takacs, J. M., Truesdale, L. K.: Tetrahedron Lett. *1977,* 2495
178. Evans, D. A., Hurst, K. M., Takacs, J. M.: J. Amer. Chem. Soc. *100,* 3467 (1978)
179. Nakamura, E., Kuwajima, J.: Angew. Chem. *88,* 539 (1976), Int. Ed. *15,* 498 (1976)
180. Peterson, D. J.: J. Org. Chem. *33,* 780 (1968)
181. Gröbel, B. T., Seebach, D.: Angew. Chem. *86,* 102 (1974); Int. Ed. *13,* 83 (1974)
182. Sakurai, H., Nishiwaki, K., Kira, M.: Tetrahedron Lett. *1973,* 4193
183. Chan, T. H., Mychajlowskij, W.: Tetrahedron Lett. *1974,* 171
184. Chan, T. H., Chang, E.: J. Org. Chem. *39,* 3264 (1974)
185. Hartzell, S. L., Sullivan, D. F., Rathke, M. W.: Tetrahedron Lett. *1974,* 1403
186. Taguchi, H., Shimoji, K., Yamamoto, H., Nozaki, H.: Bull. Chem. Soc. Jap. *47,* 2529 (1974)
187. Shimoji, K., Taguchi, H., Oshima, K., Yamamoto, H., Nozaki, H.: J. Amer. Chem. Soc. *96,* 1620 (1974)
188. Hudrlik, P. F., Peterson, D.: J. Amer. Chem. Soc. *97,* 1464 (1975)
189. Faulkner, D. J.: Synthesis *1971,* 175 (review)
190. Chan, T. H., Chang, E., Vinokur, E.: Tetrahedron Lett. *1970,* 1137
191. Corey, E. J., Enders, D., Bock, M. G.: Tetrahedron Lett. *1976,* 7
192. Sachdev, K.: Tetrahedron Lett. *1976,* 4041
193. Rühlmann, K.: Synthesis *1971,* 236
194. Schräpler, U., Rühlmann, K.: Chem. Ber. *96,* 2780 (1963)
195. Schräpler, U., Rühlmann, K.: Chem. Ber. *97,* 1383 (1964)
196. Rühlmann, K., Seefluth, H., Becker, H.: Chem. Ber. *100,* 3820 (1967)
197. Schräpler, U., Rühlmann, K.: Chem. Ber. *98,* 1352 (1965)
198. Kuwajima, J., Minami, N., Abe, T., Sato, T.: Bull. Chem. Soc. Japan *51,* 2391 (1978)
199. Rühlmann, K., Powedda, L.: J. prakt. Chem. *284,* 18 (1961)
200. Audibrand, M., Le Goaller, R., Arnaud, P.: Compt. Rend. Acad. Sci. Fr. [C] *268,* 2322 (1969)
201. Weidenhagen, R., Wegner, H.: Chem. Ber. *71,* 2124 (1938)
202. De Stereus, G., Halamandaris, A.: J. Amer. Chem. Soc. *79,* 5710 (1957)
203. Nenitzescu, C. D., Necsiou, I., Zalman, M.: Commun. Acad. Republ. Populaire romaine *7,* 421 (1957); C.A., *52,* 16330 (1958)
204. Birkofer, L., Franz, M.: Chem. Ber. *100,* 2681 (1967)
205. Birkofer, L., Franz, M.: ibid. *105,* 17 (1972)
206. Birkofer, L., Franz, M.: ibid. *104,* 3062 (1971)
207. Birkofer, L., Franz, M., ibid. *105,* 470 (1972)
208. Ford, M. F., Walton, D. R. M.: Synthesis *1973,* 47
209. Birkofer, L., Ritter, A., Richter, P.: Chem. Ber. *96,* 2750 (1963)
210. Birkofer, L., Wegner, P.: Chem. Ber. *99,* 2512 (1966)
211. Birkofer, L., Wegner, P.: Chem. Ber. *100,* 3485 (1967)
212. Tanaka, Y., Velen, S. R., Miller, S. I.: Tetrahedron *29,* 3271 (1973)

213. Birkofer, L., Stilke, R.: Chem. Ber. *107*, 3717 (1974)
214. Washburne, S. S., Peterson jr., W. R.: J. Organomet. Chem. *21*, 427 (1970)
215. Kricheldorf, H. R.: Liebigs Ann. Chem. *1973*, 1816
216. Birkofer, L., Sommer, P.: Chem. Ber. *109*, 1701 (1976)
217. Birkofer, L., Stilke, R.: J. Organomet. Chem. *74*, C 1 (1974)
218. Carboni, A., Lindsey, R. V.: J. Amer. Chem. Soc. *81*, 4342 (1959)
219. Birkofer, L., Richter, P., Ritter, A.: Chem. Ber. *93*, 2804 (1960)
220. Birkofer, L., Ritter, A., Giessler, W.: Angew. Chem. *75*, 93 (1963); Int. Ed. *2*, 96 (1963)
221. Giessler, W.: Ph. D. Thesis, Universität Köln, 1963
222. Birkofer, L., Ritter, A.: Angew. Chem. *71*, 372 (1959)
223. Birkofer, L., Kühlthau, H. P., Ritter, A.: Chem. Ber. *93*, 2810 (1960)
224. Birkofer, L., Kühlthau, H. P., Ritter, A.: Chem. Ber. *97*, 934 (1964)
225. Birkofer, L., Ritter, A., Kühlthau, H. P.: Angew. Chem. *75*, 209 (1963); Int. Ed. *2*, 155 (1963)
226. Hilbert, G. E., Johnson, T. B.: J. Amer. Chem. Soc. *52*, 4489 (1930)
227. Lukevics, E., Zablotskaya, A. E., Solomennikova, I. I.: Usp. Khim. *43*, 370 (1974) (review); Russ. Chem. Rev. (Engl. Transl.) *43*, 140 (1974); C. A., *80*, 121226 (1974)
228. Birkofer, L., Wegner, P.: Org. Synth. *50*, 107 (1970)
229. a) Washburne, S. S., Peterson jr., W. R.: J. Organomet. Chem. *33*, 156 (1971)
b) Dickopp, H., Nischk, G. E. (Farbenfabriken Bayer AG): D. O. S. 1965741; C. A., *75*, 152326 (1971)
230. Birkofer, L., Kaiser, W.: Liebigs Ann. Chem. *1975*, 266
231. McCasland, G. E.: J. Amer. Chem. Soc. *73*, 2293 (1951)
232. Birkofer, L., Kim, S. M., Ritter, A.: Chem. Ber. *96*, 3099 (1963)
233. Birkofer, L., Kim, S. M.: Chem. Ber. *97*, 2100 (1964)
234. Kricheldorf, H. R.: Synthesis *1972*, 695
235. Itoh, K., Okamura, M., Ishii, Y.: J. Organomet. Chem. *65*, 327 (1974)
236. Washburne, S. S., Peterson jr., W. R.: Synth. Commun. *2*, 227 (1972)
237. Washburne, S. S., Peterson, W. R.: J. Amer. Oil Chem. Soc. *49*, 694 (1972)
238. Kricheldorf, H. R.: Synthesis *1972*, 551
239. Kricheldorf, H. R.: Chem. Ber. *106*, 3765 (1973)
240. Peterson jr., W. R., Radell, J., Washburne, S. S.: J. Fluor. Chem. *2*, 437 (1972/73)
241. McMillan, J. H., Washburne, S. S.: J. Org. Chem. *37*, 1738 (1972)
242. Kricheldorf, H. R.: Synthesis *1974*, 561
243. Washburne, S. S., Peterson jr., W. R., Berman, D. A.: J. Org. Chem. *37*, 1738 (1972)
244. Warren, J. D., McMillan, J. H., Washburne, S. S.: J. Org. Chem. *40*, 743 (1975)
245. Kricheldorf, H. R.: Synthesis *1975*, 49
246. review: Zbiral, E.: Synthesis *1972*, 285
247. Zbiral, E., Nestler, G.: Tetrahedron *27*, 2293 (1971)
248. Ehrenfreund, J., Zbiral, E.: Tetrahedron *28*, 1697 (1972)
249. Bassindale, A. R., Brook, A. G., Jones, P. F., Stewart, J. A. G.: J. Organomet. Chem. *152*, C 25 (1978)
250. Perutz, R. N.: J. C. S. Chem. Comm. *1978*, 762
251. a) Blau, K. and King, G. S. (Ed.): "Handbook of Derivatives For Chromatography", Heyden & Sons, London, *1977*; b) Cooper, B. E.: Chem. Ind. (London) *1978*, 794
252. Kowalski, J., Lasocki, Z.: J. Organomet. Chem. *128*, 37 (1977)
253. Yoder, C. H., Copenhafer, W. C., du Beshter, B.: J. Amer. Chem. Soc. *96*, 4283 (1974)
254. Kowalski, J., Lasocki, Z.: J. Organomet. Chem. *116*, 75 (1976)
255. Hillyard, jr., R. W., Ryan, C. M., Yoder, C. H.: J. Organomet. Chem. *153*, 369 (1978)
256. Lasocki, Z., Kowalski, J.: J. Organomet. Chem. *152*, 45 (1978)
257. Jancke, H., Engelhardt, G., Rühlmann, K. et al.: J. Organomet. Chem. *134*, 21 (1977)
258. Birkofer, L., Donike, M.: J. Chromatogr. *26*, 270 (1967)
259. Birkofer, L., Müller, F.: in "Peptides 1968" (Lectures Internat. Peptide Symp. Paris 1968), p. 151, Amsterdam: North-Holland 1968
260. Birkofer, L., Konkol, W., Ritter, A.: Chem. Ber. *94*, 1263 (1961)

261. Birkofer, L., Ritter, A., Neuhausen, P.: Liebigs Ann. Chem. *659*, 190 (1962)
262. Birkofer, L., Schramm, J.: Liebigs. Ann. Chem. *1975*, 2195
263. Birkofer, L., Schramm, J.: Chem. Ber. *95*, 426 (1962)
264. Birkofer, L., Schramm, J.: Liebigs Ann. Chem. *1977*, 760
265. Krüger, C., Rochow, E. G., Wannagat, U.: Chem. Ber. *96*, 2132 (1963)
266. Pinnick, H. W., Bal, B. S., Lajis, N. H.: Tetrahedron Lett. *1978*, 4261
267. Corey, E. J., Snider, B. B.: J. Amer. Chem. Soc. *94*, 2549 (1972)
268. Negishi, E., Law, G., Yoshida, T.: J. C. S. Chem. Comm. *1973*, 874
269. Weis, R., Pfaender, P.: Liebigs Ann. Chem. *1973*, 1269
270. Corey, E. J., Venkateswarlu, A.: J. Amer. Chem. Soc. *94*, 6190 (1972)
271. Corey, E. J., Sachdev, H. S.: J. Amer. Chem. Soc. *95*, 8483 (1973)
272. Prestwich, G. D., Labowitz, J. N.: J. Amer. Chem. Soc. *96*, 7103 (1974)
273. Yankee, E. W., Axen, U., Bundy, G. L.: J. Amer. Chem. Soc. *96*, 5865 (1974)
274. Marshall, J. A., Peveler, R. D.: Synth. Commun. *3*, 167 (1973)
275. Ogilvie, K. K., Iwacha, D. J.: Tetrahedron Lett. *1973*, 317
276. Ganem, B., Small jr., V. R.: J. Org. Chem. *39*, 3728 (1974)
277. Brandstetter, H. H., Zbiral, E.: Helv. Chim. Acta *61*, 1832 (1978)
278. Franke, F., Guthrie, R. D.: Austral. J. Chem. *30*, 639 (1977)
279. Ogilvie, W. R.: Canad. J. Chem. *51*, 3799 (1973)
280. Ogilvie, W. R., Sadana, K. L., Thompson, E. A., Westmore, J. B.: Tetrahedron Lett. *1974*, 2861
281. Franke, F., Guthrie, R. D.: Austral. J. Chem. *31*, 1285 (1978)
282. Ogilvie, W. R., Sadana, K. L., Thompson, E. A., Westmore, J. B.: Tetrahedron Lett. *1974*, 2865
283. Corey, E. J., Ravindranathan, T.: J. Amer. Chem. Soc. *94*, 4013 (1972)
284. Birkofer, L., Krebs, K.: Tetrahedron Lett. *1967*, 885
285. Birkofer, L., Sommer, P.: J. Organomet. Chem. *35*, C 15 (1972)
286. Birkofer, L., Sommer, P.: J. Organomet. Chem. *99*, C 1 (1975)
287. Birkofer, L., Havix, J.: in press
288. Knausz, D. et al.: Vth Internat. Symp. Organosilicon Chem., Karlsruhe 1978, abstracts, p. 48
289. Birkofer, L.: unpublished results
290. Brewer, St. D., Haber, Ch. F.: J. Amer. Chem. Soc. *70*, 3888 (1948)
291. Stuhl, O., Ph. D. Thesis, Universität Düsseldorf 1978
292. Kelly, R. W.: Tetrahedron Lett. *1969*, 967
293. Findlay, J. K., Siekmann, L., Breuer, H.: Biochem. J. *137*, 263 (1974)
294. Kelly, R. W.: J. Chromatogr. *43*, 229 (1969)
295. Wieber, M., Schmidt, M.: Chem. Ber. *96*, 1561 (1963)
296. Mehrotra, R. C., Narain, R. P.: Indian J. Chem. *5*, 444 (1967)
297. Birkofer, L., Stuhl, O.: J. Organomet. Chem. *164*, C 1 (1979)
298. a) Birkofer, L., Stuhl, O.: J. Organomet. Chem. in press
b) Birkofer, L., Stuhl, O.: J. Organomet. Chem. *177*, C16 (1979)
299. Birkofer, L., Kraemer, E.: Chem. Ber. *100*, 2776 (1967)
300. Pray, B. O., Sommer, L. H., Goldberg, G. M., Kerr, G. T., Di Giorgio, P. A., Whitmore, F. C.: J. Amer. Chem. Soc. *70*, 433 (1948)
301. Krueerke, U.: Chem. Ber. *95*, 174 (1962)
302. Ho, Tse-Lok, Olah, G. A.: Angew. Chem. *88*, 845 (1976); Int. Ed. *15*, 774 (1976)
303. Ho, Tse-Lok, Olah, G. A.: Synthesis *1977*, 417
304. Jung, M. E., Ornstein, P. J.: Tetrahedron Lett. *1977*, 2659
305. Jung, M. E., Blumenkopf, T. A.: Tetrahedron Lett. *1978*, 3657
306. McKenna, C. E., Higa, M. T., Cheung, N. H., McKenna, M. C.: Tetrahedron Lett. *1977*, 155
307. Gross, H., Boeck, C., Costisella, B., Gloede, J.: J. prakt. Chem. *320*, 344 (1978)
308. Evans, D. A., Hoffmann, J. M., Truesdale, L. K.: J. Amer. Chem. Soc. *95*, 5822 (1973)
309. Zubrick, J. W., Dunbar, B. J., Durst, H. D.: Tetrahedron Lett. *1975*, 71

310. Ryu, I., Murai, S., Horiike, T., Shinonaga, A., Sonoda, N.: Synthesis *1978*, 154
311. Uznanski, B., Stec, W. J.: ibid. *1978*, 154
312. Rasmussen, J. K., Heilmann, S. M.: ibid. *1978*, 219
313. Evans, D. A., Truesdale, L. K., Carroll, G. L.: J. C. S. Chem. Commun. *1973*, 55
314. Neek, H., Müller, R.: J. prakt. Chem. *315*, 367 (1973)
315. see also: Hünig, S., Wehner, G.: Synthesis *1975*, 180
316. Lidy, W., Sundermeyer, W.: Chem. Ber. *106*, 587 (1973)
317. Lidy, W., Sundermeyer, W.: Tetrahedron Lett. *1973*, 1449
318. Evans, D. A., Hoffman, J. M.: J. Amer. Chem. Soc. *98*, 1983 (1976)
319. Evans, D. A., Carroll, G. L., Truesdale, L. K.: J. Org. Chem. *39*, 914 (1974)
320. Parham, E., Roosevelt, C. S.: Tetrahedron Lett. *1971*, 923
321. Ryu, I., Murai, S., Shinonaga, A., Horiike, T., Sonoda, N.: J. Org. Chem. *43*, 780 (1978)
322. Deuchert, K., Hertenstein, U., Hünig, S.: Synthesis *1973*, 777
323. Evans, D. A., Wong, R. Y.: J. Org. Chem. *42*, 350 (1977)

Received July 5, 1979

The 4a,4b-Dihydrophenanthrenes

Karol A. Muszkat

Department of Structural Chemistry, The Weizmann Institute of Science, Rehovot, Israel

Table of Contents

I Introduction

4a,4b-Dihydrophenanthrenes are colored unstable conjugated polyenes obtained photochemically by the irradiation of the corresponding cis-1,2-diarylethylenes. Thus 4a,4b-dihydrophenanthrenes (*1*), the parent molecule[1], is formed from optically excited cis-stilbene,

cis-stilbene —UV LIGHT→ 4a, 4b - dihydrophenanthrene *1*   A

As in many other instances, 4a,4b-dihydrophenanthrenes were purposedly studied only in recent years even though their formation was unknowingly observed many years ago. Thus, Lewis, Magel and Lipkin noted already in their pioneering 1940 paper[1)] that

"... *whenever a solution containing cis-stilbene is irradiated a yellow substance is gradually formed*".

This yellow coloration[2] is undoubtedly due to DHP (*1*), whose oxidation product, phenanthrene, was observed by Smakula as an oxidation product in the photolysis of cis-stilbene:[3)]

"... *auch cis-Stilben nicht beständig ist. Es geht aber bei der Bestrahlung nicht in trans-Stilben über, sondern in einen Stoff mit einem Absorptionsmaximum bei 247 mμ.*"

Only much later could Parker and Spoerri relate the 247 mμ absorption maximum to the ^1B_b band of phenanthrene[4)]. The observations of Lewis, Magel and Lipkin of Smakula and the finding of Parker and Spoerri were finally correlated in 1963 by Moore, Morgan and Stermitz[5)], who suggested the 4a,4b-dihydrophanthrene structure (*1*) for the yellow intermediate of Lewis et al. and reported on the thermal and photochemical ring opening processes,

(photochemical) DHP $\xrightarrow{h\nu}$ cis-stilbene   B

(thermal) DHP $\xrightarrow{\Delta}$ cis-stilbene   C

During the past two decades the photocyclization of substituted diaryl ethylenes (analogous to A) has assumed considerable synthetic importance. In such cases photocyclization is carried out simultaneously with dehydrogenation (e.g., by iodine), the aim being the one step formation of the fully aromatic system[6)].

1 We shall design 4a,4b-dihydrophenanthrene as DHP

2 The yellow coloration reported by Ciamician and Silber[2)] when trans-stilbene solutions were exposed to sunlight seems probably due to stable oxidation products and not to DHP itself

DHP (−2H) → phenanthrene D

As we shall not be concerned directly with purely synthetic aspects of the photocyclization we would like to refer the interested reader to the comprehensive general reviews of Stermitz[7], Blackburn and Timmons[8] and to the yearly reviews of Gilbert[9].

However, before leaving this subject we would like to single out the remarkable synthetic activity in the helicenes series. In this rapidly developing field truly outstanding progress has been made possible by the photocyclodehydrogenation process introduced for this purpose by R. H. Martin et al.[10].

Besides their obvious role as reactive intermediates in a powerful synthetic approach the 4a,4b-dihydrophenanthrenes offer a fascinating combination of unusual chemical and physical properties. Over the past 15 years these topics were investigated at length at the Weizmann Institute in Rehovot and elsewhere, and the present review is intended to provide an up-to-date summary of the activity in this field.

II Survey of Known Systems

Tables 1–9 list all the 4a,4b-dihydrophenanthrenes and their analogs that have been observed (to the best of our knowledge) up to the time of the writing of this review. Several of the compounds described in Tables 1–9 (e.g. *6, 15–19, 21,* and *60*) were studied on various occasions in Rehovot but were not included in previous publications. The data on several other systems listed in these Tables are published for the first time though the compounds themselves were mentioned in previous reports.

An important point we wish to stress within the present context is that the number of observed 4a,4b-dihydrophenanthrenes is far smaller than the number of systems in which the photocyclodehydrogenation process (e.g. A. followed by D.) has been reported. In many cases the reason is simply that these intermediates were not looked for so that no special efforts were made to observe them. However, in many instances in which photocyclodehydrogenation products are known to be formed no 4a,4b-dihydrophenanthrenes can be observed even under usually favorable conditions (see below). In this case either the 4a,4b-dihydrophenanthrenes are destroyed by some subsequent process or that the photostationary concentration of these species is too low. Low photostationary concentrations are due (among other causes, see below) to low cyclization quantum yields. Such is the case, e.g., with stilbenes substituted at the 4-ring position with electron attracting groups.

4a,4b-Dihydrophenanthrene systems are formed from cis-1,2-diphenylethylene or from more complex systems which contain a cis-1,2-diphenylethylene subunit. On this basis the molecules listed in Tables 1–9 are subdivided into the following groups:

Table 1. 4a,4b-Dihydrophenanthrenes. Systems derived from Stilbenes

1	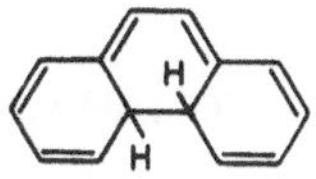 [structure]	from Stilbene (5×10^{-4} M, MCH/IH), UV[a], λ_i 280 c 22% at −30° $\lambda_{max}(\epsilon)$I:450(6,750) II:310 (22,200) 297(20,700), III:237 (15,200). ϕ_c ~ 0.1, ϕ_o 0.3, (λ_i 313) ϕ_o 0.7 (λ_i 436), tr 10–40°, E_a 17.5, τ_{25} 96 min, E_c 1.2 Ref.[11])
2	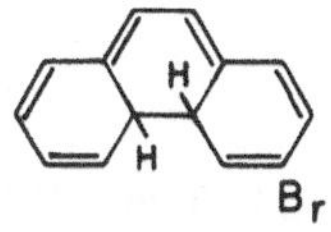 [structure, Br]	from 4-bromostilbene λ_i 280 λ_{max} 450; D 0.65 (3.8×10^{-4} M) at 0° tr 25–40°; τ_{25} 106 min; E_a 23, Ref.[11])
3	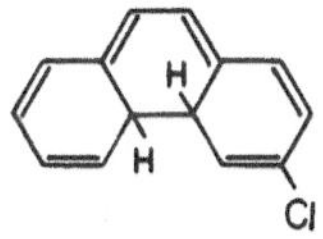 [structure, Cl]	from 4-chloro stilbene, 4.6×10^{-4} M, MCH, λ_i 280 at +20° λ_{max} 450; D 0.35 tr 20–40°; τ_{25} 90 min; E_a 19, Ref.[11])
4	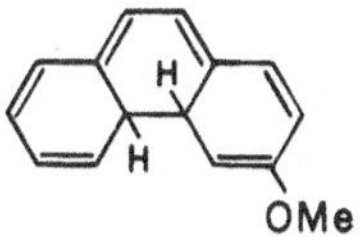 [structure, OMe]	from 4-methoxystilbene, 5×10^{-4} M, MCH, at −30°, UV[a] λ_{max} 454 D 0.05. *
5	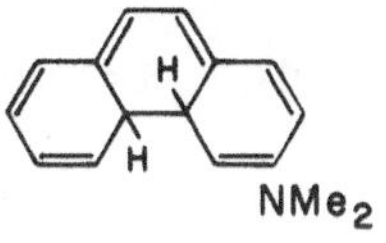 [structure, NMe$_2$]	from 4-dimethylaminostilbene in MCH λ_{max} 450. *
6	 [structure, NC CN]	two modifications have been obtained by irrad. (UV[a]) of α,α'-dicyanostilbene in MCH: '475' is formed exclusively above −40°. In 2×10^{-4} M sol. λ_{max} 475; D 0.30, τ_{-40} 9 min. '510' predominates at −80 °C λ_{max} 510; D 0.27 τ_{-40} 57 min. At −80°, '475' is formed first, followed by '510'.*, Ref.[12a, b])
7	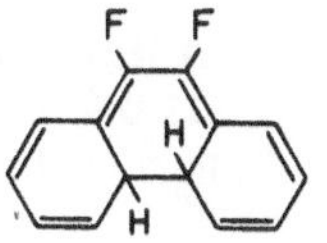 [structure, F F]	from α,α-difluorostilbene, 4.6×10^{-4} M, MCH. λ_i 280 at −10°. λ_{max} 430; D 0.07, Ref.[11])
8	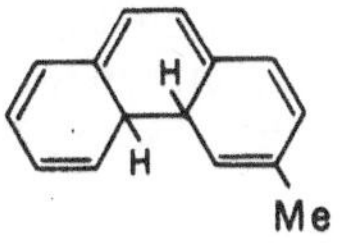 [structure, Me]	from 4-methylstilbene, 4.1×10^{-3} M, IO, UV[a], at −30° λ_{max} 447 D 1.15. *

Table 1 (continued)

9	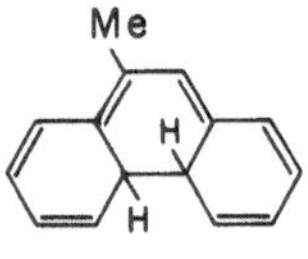 from α-methylstilbene, 1.3×10^{-3} M, IO, UV[a] λ_{max} 452; D 0.10. Ref.[13] *
10	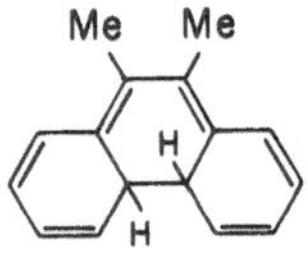 from α,α′-dimethylstilbene, 10^{-3} M, MCH at $-30°$, UV[a], λ_{max} 455; D 0.09, Ref.[13] *
11	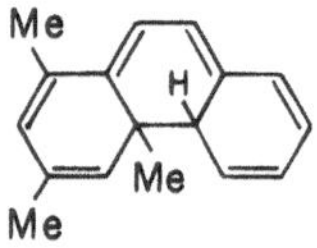 from 2,4,6-trimethylstilbene, 8.8×10^{-4} M, MCH at 0°, λ_{max} 460; D 0.35. Ref.[11, 13]*
12	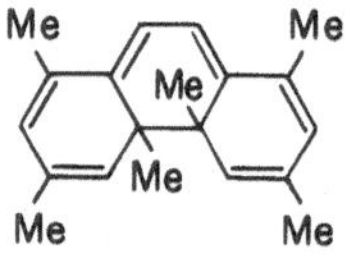 from 2,2′,4,4′,6,6′-hexamethylstilbene, 4×10^{-4}M, MCH/IH, λ_i 280 c 21% at +10° λ_{max} (ϵ)I: 475 (3100) II: 320 (7100), 310 (7000) III: 245 (15800). tr. 40–75°, E_a 22.5, τ_{25} 33 h ϕ_c 0.04, ϕ_o 0.4 (280). Ref.[11, 14]
13	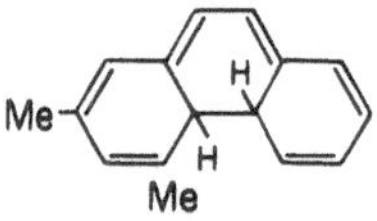 from 3,5-dimethylstilbene, 3×10^{-3} M, IO, at $-31°$, UV[a] λ_{max}(D) 458 (1.26)Sh; 434 (1.52) 410 (1.15)Sh, $\tau_{-31°}$ 2 h, τ_{-10} 10 min, Ref.[13]*
14	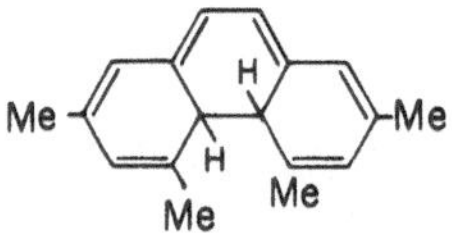 from 3,5,3′,5′-tetramethylstilbene in MCH/IH UV[a] at $-180°$ c 20%. In MCH at $-120°$ λ_{max} 525 D 0.245 τ_{-120} 6 min; τ_{-160} = 43 min Ref.[13], E_a 7 *[89]
15	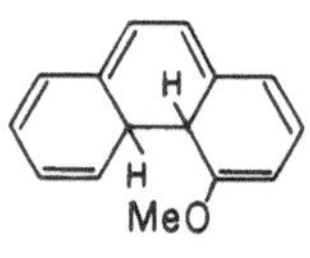 from m-methoxy stilbene, 5×10^{-4} M, MCH/MCP $-30°$, λ_i: UV[a] λ_{max}(D): 485 (0.58); 458 (0.74); 430 (0.55) measured at $-180°$. Ring opening with λ 436 at $-180°$. * Ref.[15]

Table 1 (continued)

16	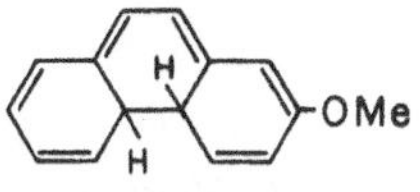 see 4 methoxy-DHP. * Ref.[15]
17	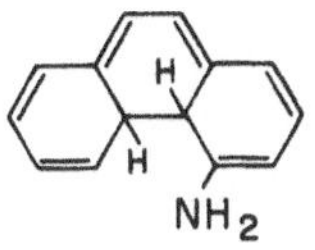 see 2-amino-DHP. * Ref.[15]
18	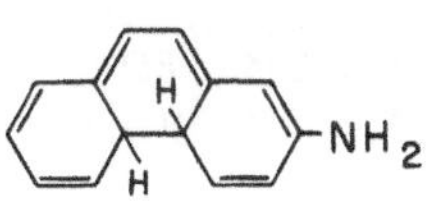 from m-aminostilbene, 5×10^{-4} M, MTHF, λ_i UV[a] at $-30°$. Stable isomer λ_{max} 497 D 0.92. Unstable isomer λ_{max} 510. Ring opening with λ_i 436 at $-30°$ and $-180°$. Oxidises rapidly with O_2 at $-30°$. Moderately stable at $-30°$. * Ref.[15]
19	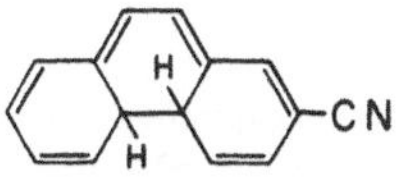 from m-cyanostilbene, 6.6×10^{-4} M, MCH, λ_i 280 at $-30°$. λ_{max}(D)467 (0.26) λ_{max} 442 (0.26). * Ref.[15]
20	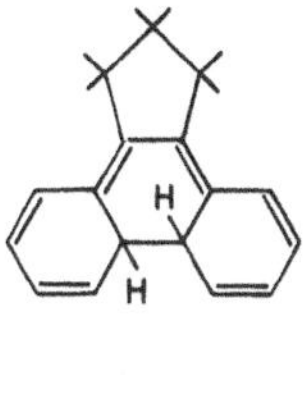 from 1,2-diphenylcyclopentene, 5.4×10^{-4} M MCH/IH, λ_i 280 at $-20°$; c 67%; $\lambda_{max}(\epsilon)$ I: 460 (7500) II: 320 (22,200); 306 (19,500), III: 246 (15,000). tr 10–40°, E_a 15.5, τ_{25} 23 min $\phi_c(\lambda)$ 0.43 (313), ϕ_o 0.48 (436), both at $-20°$. E_c 2.5, Ref.[11]
21	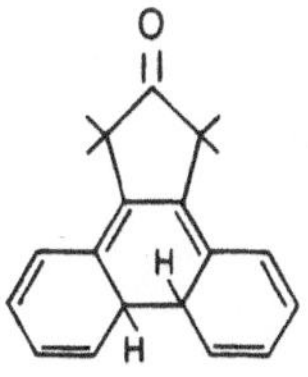 from 1,2-diphenylcyclopentenone, 2.1×10^{-4} M, MCH/IH, $-30°$, λ_i 280 λ_{max}472 D 0.77 *
22	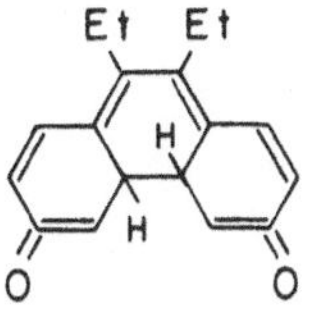 from stilbesterol, 3.7×10^{-5} M, E, λ_i 254 λ_{max}(D), I: 406 (19600); II: 287 (23600), III: 221 (8900). UV[a], λ_{max} 412 D 0.45 Ref.[16] *

Table 2. 4a,4b-Dihydrophenanthrenes bridged at position 4 and 5, derived from [2.2]metacyclophanes

23	from [2.2] metacyclophanene 3×10^{-5} M, MCH/MCP at 0°; λ_i 254 c 80%, $\lambda_{max}(\epsilon)$ I: 500 (3,500); II: 319 (18,400); 306 (15,200); 293 (7.800); ϕ_c 0.5(λ_i 254), ϕ_o 0.58 (λ_i 546) tr 30–60°, E_a 20.3; τ_{25} 61 h, Ref.[17, 18]
24	from 4-methyl-[2.2]metacyclophanene, 3×10^{-5} M, MCH/MCP at 0°, λ_i 254, c 45%, $\lambda_{max}(\epsilon)$ I: 500 (3.200); II: 322 (17,000); 308 (14,100); 294 (6.850); ϕ_c 0.53 (λ_i 254), ϕ_o 0.63 (λ_i 546) tr 30–60°, E_a 19; τ_{25} 56.9 h, Ref.[17, 18]
25	from 4,12-dimethyl-[2.2]metacyclophanene, 3×10^{-5} M, MCH/MCP at 0°, λ_i 254, c 80% $\lambda_{max}(\epsilon)$ I: 520 (3,000) II: 323 (16,000); 309 (13,100); 296 (6,900). ϕ_c 0.65 (λ_i 313, −50°), ϕ_o 0.39 (λ_i 546, −50°). tr 30–50°, E_a 15, τ_{25} 916 h, Ref.[17, 18]
26	from tetramethylaza[2.2]metacyclophanene, CH, by UV irrad. Deep red, λ_{max} 507. Fades in 2 min at RT. Ref.[19]

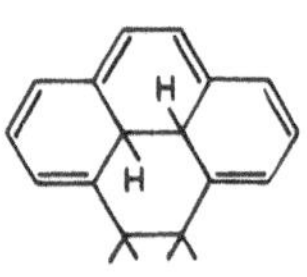

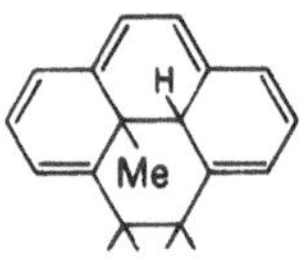

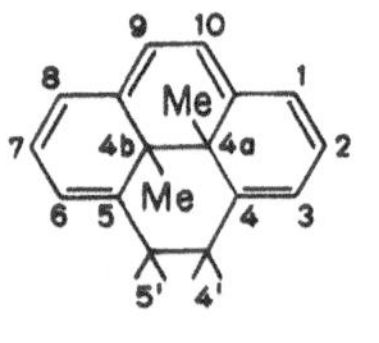

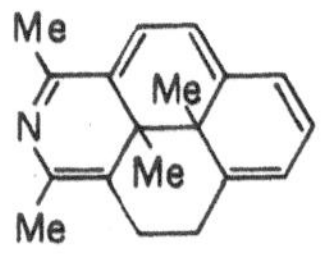

Table 3. 4a,4b-Dihydrophenanthrenes. Heterocyclic analogs with three conjugated rings

	Structure	Description
27	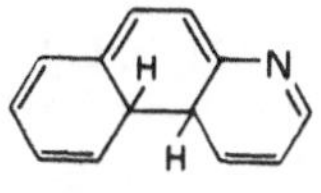 H, N, H	from 2-stilbazole, 5×10^{-4} M, MCH/IH, UV[a] at 0° λ 460 D 0.12. * Ref.[47]
28	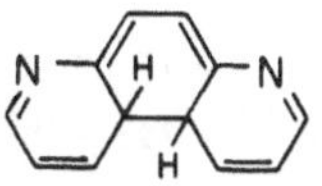 N, H, N, H	from 1,2-di(2-pyridyl)ethylene, 7×10^{-4} M, MCH, UV[a], λ_{max} 450, D 0.37 at 0°, at −30°, D 0.46, at −50°, D 0.76; at −100°, D 0.97; at −140° two isomers λ_{max} 450 and λ_{max} 470. * Ref.[20]
29	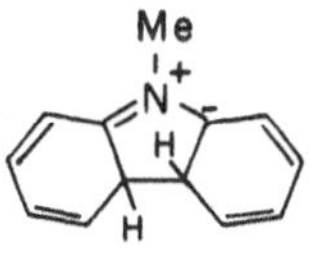 Me, N^+, H, H	from N-methyldiphenylamine 10^{-4} M, MCH, UV[a], $\lambda_{max}(\epsilon)$ I: 610 (21,000), II: 375 (3,000) tr −30° to 25°, E_a 17, $\tau_{-30°}$ 1 s, $\tau_{25°}$ 40 ms E_c 5.5 Kcal/mole. Ref.[21]
30	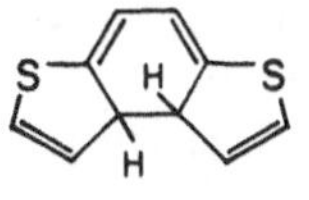 S, H, S, H	from 1,2-di(2-thienyl)ethylene, 0.005 M, CH, UV[b], λ_{max} ca. 360; ϕ_c 0.07; fairly stable at RT for 12–15 h, Ref.[22]*
31	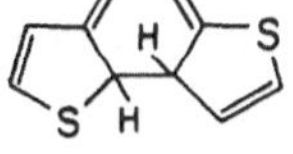 H, S, S, H	from 1,2-di(2,3′-thienyl)ethylene, 0.005 M, CH, UV[b], UV[a]: λ_{max} 390; D 0.3, reasonably stable at RT. Erased by λ_i 436. Ref.[22] *
32	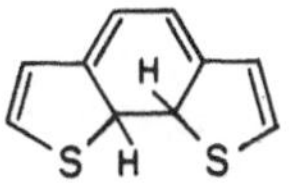 H, S, H, S	from 1,2-di(3-thienyl)ethylene 0.005 M, CH, UV[b]. λ_{max} 420, D 0.26, also by UV[a] in MTHF D 0.46, erased by λ_i 436, same in T. Ref.[22] *
33	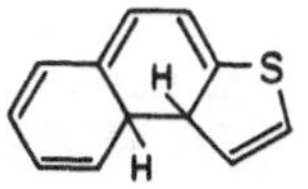 H, S, H	from 1-phenyl 2-(2-thienyl)ethylene, 0.005 M, CH, UV[b], λ_{max} 415, D 0.14, Ref.[22] *

Table 4. 4a,4b-Dihydrophenanthrenes. Systems with four conjugated rings derived from 1-phenyl-2-naphtylethylenes

34	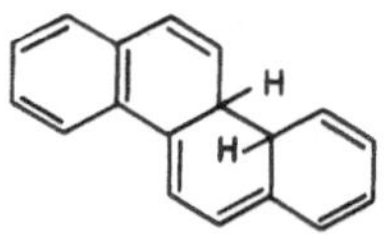	from 1-styrylnaphthalene, 2×10^{-4} M, H, UV[a], λ_{max} 423.7 (ϵ 8000). In MCH/IH (10^{-5} M) λ_i 334 at $-10°$, c 50%. tr 20–50°, E_a 12.8; τ_{25} 3.57 h, ϕ_c 0.3 (λ_i 334); ϕ_o 0.7 (λ_i 436) E_c 5.5 Ref.[23–25)]
35	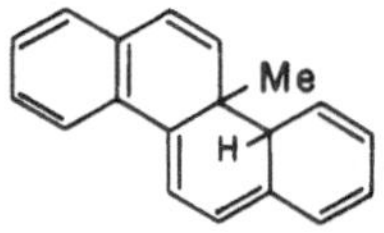	from 1-styryl 2-methyl naphtalene, 2×10^{-4} M, H, UV[a], λ_{max} 436.7. tr 20–40°, E_a 15.6; τ_{25} 2.03 h, Ref.[23)]
36	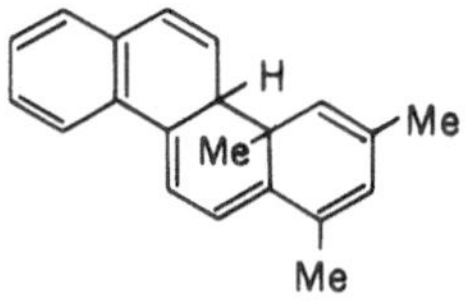	from 1-(2,4,6-trimethyl styryl)naphtalene, 2×10^{-4} M, H, UV[a], λ_{max} 434.7 tr 20–45°, E_a 23.5; τ_{25} 55 h, Ref.[23)]
37	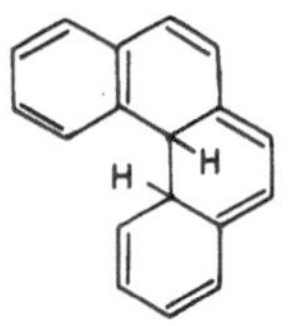	(ϕ-2N)$_A$ from 2-styrylnaphtalene, 2×10^{-4} M, H, UV[a]; λ_{max} 446.4, 421.9 (ϵ 11,000), 401.6 at $-185°$, 10^{-5} M in MCH/IH, λ_i 334 – c 25%. tr 25–45°, E_a 14.2; τ_{25} 1 h, ϕ_c 0.2 (λ_i 334, air, 0°); ϕ_o 0.7 (λ_i 436); ϕ_F 0.05 ($-180°$) E_c 4.5, E_o 3. Ref.[23–25)]
38	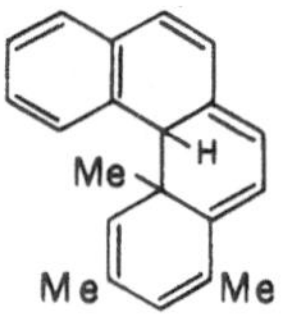	from 2-(2,4,6-trimethylstyryl)naphthalene 2×10^{-4} M, H, UV[a]; λ_{max} 487.8, 460.8, 442.4, 413.2. tr 20–40°; E_a 20.7; τ_{25} 39.8 min, Ref.[23)]
39	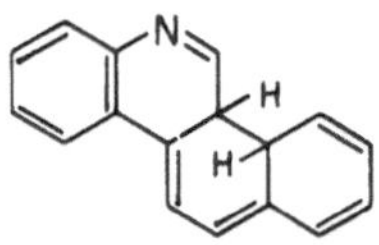	from 2-styrylquinoline, 2×10^{-4} M, H, UV[a]; λ_{max} 437; tr 20–40°; E_a 7.3; τ_{25} 8.7 min, Ref.[23)]
40	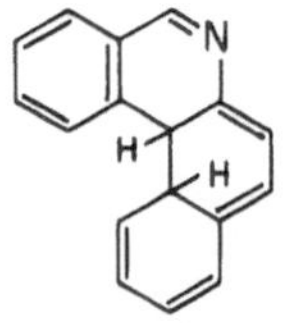	from 3-styrylisoquinoline 2×10^{-4} M, H, UV[a], λ_{max} 454.5, 438.6, 421.9, 403.2, tr 20–45°; E_a 15.7; τ_{25} 102 min, Ref.[23)]

Table 5. 4a,4b-Dihydrophenanthrenes. Systems with five conjugated rings derived from 1,2-dinaphthylethylenes

41	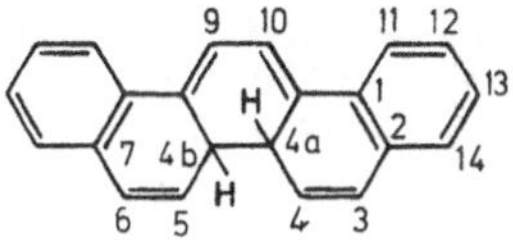 9 10 11 12 13 14 1 2 3 4 4a 4b 5 6 7 H H	from 1,2-di(1-naphthyl)ethylene, 10^{-5} M, MCH/IH, λ_i 334, λ_{max} 410, ϵ 13,000, c 50%, tr 75–100°, E_a 15, τ 11 days, ϕ_c 0.23 (air, λ_i 334, 0°), ϕ_o 0.5 (λ_i 436, 0°) E_c 2.5, Ref.[11, 24, 25]
42	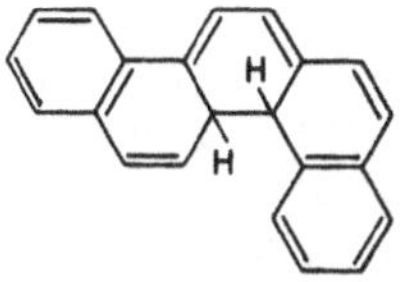 H H	$(1N\text{-}2N)_A$ from 1-(1-naphthyl)-2-(naphthyl)ethylene, 10^{-5} M, MCH/IH, λ_i 334, c 70% (−160°) λ_{max} 410 ϵ 13,000 tr 75–100°, E_a 18.5, τ 3 days ϕ_c 0.06 (λ_i 334), ϕ_o 1. (λ_i 436), ϕ_F 0.3 (−180°), E_c 2.5, E_o 4, Ref.[24,25]
43	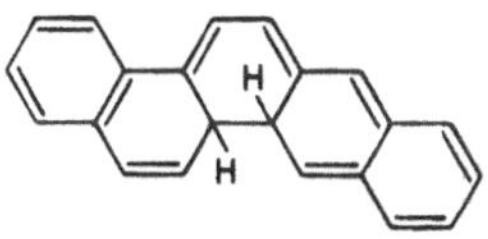 H H	$(1N\text{-}2N)_B$, from 1-(1-naphthyl)-2-(2-naphthyl)ethylene, λ_{max} 603, 570, 540, tr 20–70°, E_a 11.5, τ_{20} 46 ms, ϕ_c 0.003 (λ_i 313), Ref.[24, 25]
44	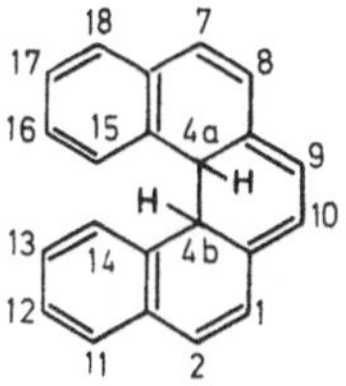 18 7 17 8 16 15 4a 9 H H 10 13 14 4b 12 1 11 2	$(2N\text{-}2N)_A$, from 1,2-di(2-naphthyl)ethylene, λ_i 365 λ_i 365 RT c 45% H or MCH/IH λ_{max} 448.4, 421.9, (ϵ 12,000), 396.8, tr 25–50°, E_a 23.3; τ_{25} 36.8 days ϕ_c 0.02, ϕ_o 0.008, ϕ_F 0.45 at −160°, ϕ_F 0.7 E_c 10, E_o 6.5, Ref.[23–27]
45	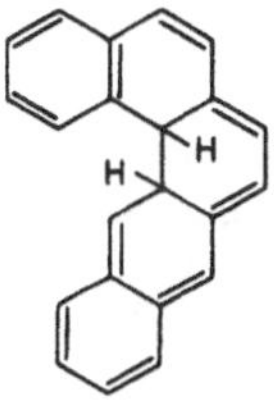 H H	$(2N\text{-}2N)_B$, from 1,2-di(2-naphthyl)ethylene, λ_{max} 572, (ϵ 10,000) 530, 495, E_a 12, $\tau_{-140°}$ 1 h, λ_i 313 ϕ_c 0.06 (−50°), ϕ_o 10^{-3} (−160°), Ref.[24–27]
46	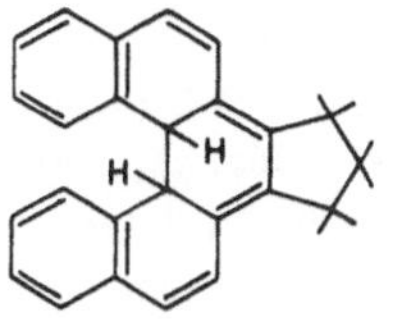 H H	$(DNCP)_A$, from 1,2-di(2-naphthyl)cyclopentene, 10^{-5} M, MCH/IH λ_i 365, c 40%(RT) λ_{max} I: 466, 440 (ϵ 7,500), 415 tr 70–95° E_a 29 τ 7 days ϕ_c 0.17 (λ_i 366); ϕ_o 0.03 (λ_i 436) ϕ_F 0.60 E_c 9, E_o 13.7, Ref.[24b]
47	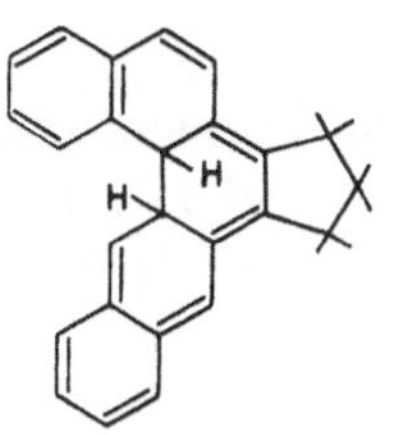 H H	$(DNCP)_B$, from 1,2-di(2-naphthyl)cyclopentene, λ_{max} 585 (ϵ 6,400) tr − 70 to 0° E_a 10 τ_{25} 5 x 10^{-4}s ϕ_c 0.14 (λ_i 313), Ref.[24b]

Table 6. 4a,4b-Dihydrophenanthrenes with 6–9 conjugated rings, derived from 1,2-diarylethylenes

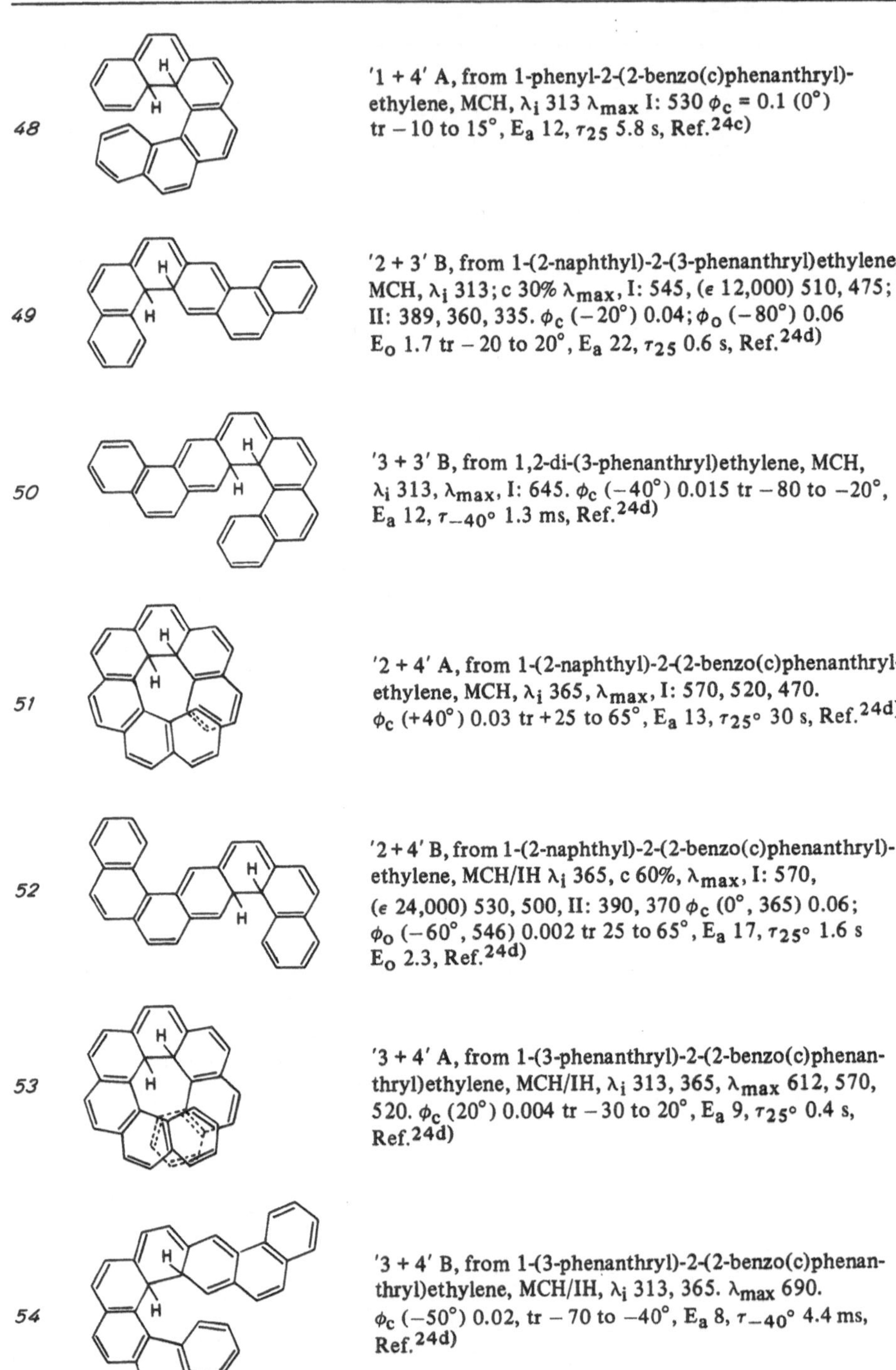

48	'1 + 4' A, from 1-phenyl-2-(2-benzo(c)phenanthryl)-ethylene, MCH, λ_i 313 λ_{max} I: 530 $\phi_c = 0.1$ (0°) tr – 10 to 15°, E_a 12, τ_{25} 5.8 s, Ref.[24c)]
49	'2 + 3' B, from 1-(2-naphthyl)-2-(3-phenanthryl)ethylene, MCH, λ_i 313; c 30% λ_{max}, I: 545, (ϵ 12,000) 510, 475; II: 389, 360, 335. ϕ_c (–20°) 0.04; ϕ_o (–80°) 0.06 E_o 1.7 tr – 20 to 20°, E_a 22, τ_{25} 0.6 s, Ref.[24d)]
50	'3 + 3' B, from 1,2-di-(3-phenanthryl)ethylene, MCH, λ_i 313, λ_{max}, I: 645. ϕ_c (–40°) 0.015 tr – 80 to –20°, E_a 12, $\tau_{-40°}$ 1.3 ms, Ref.[24d)]
51	'2 + 4' A, from 1-(2-naphthyl)-2-(2-benzo(c)phenanthryl-ethylene, MCH, λ_i 365, λ_{max}, I: 570, 520, 470. ϕ_c (+40°) 0.03 tr + 25 to 65°, E_a 13, $\tau_{25°}$ 30 s, Ref.[24d)]
52	'2 + 4' B, from 1-(2-naphthyl)-2-(2-benzo(c)phenanthryl)-ethylene, MCH/IH λ_i 365, c 60%, λ_{max}, I: 570, (ϵ 24,000) 530, 500, II: 390, 370 ϕ_c (0°, 365) 0.06; ϕ_o (–60°, 546) 0.002 tr 25 to 65°, E_a 17, $\tau_{25°}$ 1.6 s E_o 2.3, Ref.[24d)]
53	'3 + 4' A, from 1-(3-phenanthryl)-2-(2-benzo(c)phenanthryl)ethylene, MCH/IH, λ_i 313, 365, λ_{max} 612, 570, 520. ϕ_c (20°) 0.004 tr – 30 to 20°, E_a 9, $\tau_{25°}$ 0.4 s, Ref.[24d)]
54	'3 + 4' B, from 1-(3-phenanthryl)-2-(2-benzo(c)phenanthryl)ethylene, MCH/IH, λ_i 313, 365. λ_{max} 690. ϕ_c (–50°) 0.02, tr – 70 to –40°, E_a 8, $\tau_{-40°}$ 4.4 ms, Ref.[24d)]

Table 6 (continued)

55	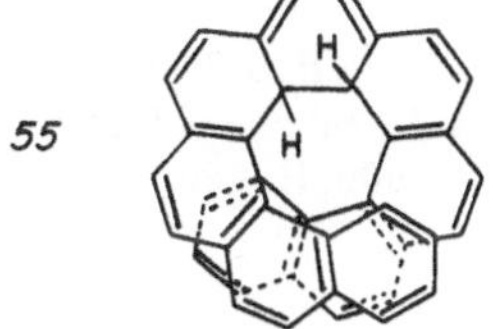	'4 + 4' A, from 1,2-bis-(2-benzo(c)phenanthryl)ethylene, MCH, λ_i 365 (20°) λ_{max} 560. ϕ_c (20°) 0.002 tr −40 to 20°, E_a 8, $\tau_{25°}$ 1.4 s, Ref.[24d)]

Table 7. 4a,4b-Dihydrophenanthrenes with 7 conjugated rings, derived from 10,10'-DiH-dianthrylidenes

56	C photoisomer, from 10,10'-di H-bianthrylidene, in MCH/IH at −100°, λ_i 313 c 75% $\lambda_{max}(\epsilon)$ I: 500 (3,360), II: 380 (17,700), 363 (15,850). E_a 12.6; τ_{-20} 0.92 s ϕ_c (−110°) 0.30, Ref.[28)]
57	C photoisomer from 10,10'-di H-10,10'-dihydroxybianthrylidene in 1-P/2-P, 4.9 x 10^{-5} M, at −100°, λ_i 313, λ_{max} I: 505, II: 377, 358 c 58%, tr −35 to 21°; E_a 14.7; τ_{-20} 1.6 s, ϕ_c (−80°) 0.60, Ref.[28)]
58	C photoisomer from 10,10'-di H-10,10-dihydroxy-1,1'-dimethyl bianthrylidene in 1-P/2-P at −130 °, λ_i 313, λ_{max} I: 560, II: 370, 355 tr −81 to 21°, E_a 15; τ_{-20} 0.52 s, ϕ_c (0°) 0.51 ϕ_o (−100°) 0.3, Ref.[28)]

Table 8. 4a,4b-Dihydrophenanthrenes with 7 conjugated rings, derived from bianthrones

59	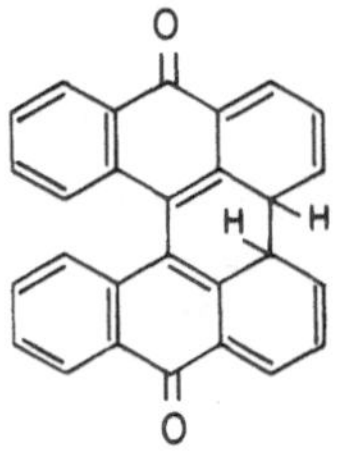 [structure]	C photoisomer, observed in flash-photolysis of biantrhone, 5×10^{-5} M, 2-P at $-75°$ also in A. In 2-P $\tau_{-75°}$ 0.08 s giving dihydrohelianthrone. λ_{max} I: 600, II: 460, 433. Ref.[29)]
60	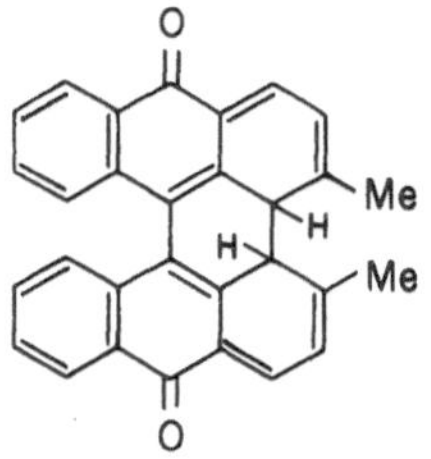 [structure]	C photoisomer, from 2,2′-dimethylbianthrone, 5×10^{-5} M, E, λ_i 366 + 405, at $-160°$. λ_{max}, I: 600, II: 460, 430. *
61	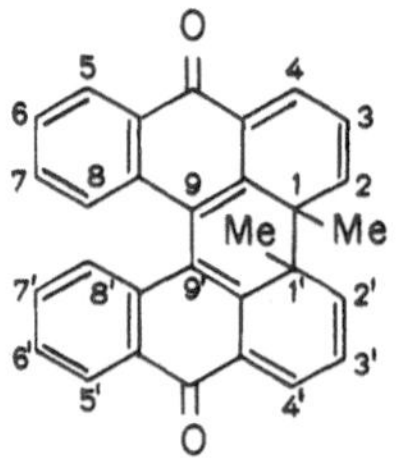 [structure]	C photoisomer, from 1,1′-dimethylbianthrone, 4×10^{-3} M, MC, $-90°$, λ_i 405(0.1 mm path) λ_{max} I: 650, II: 460, 480 c 90%. ϕ_c 0.5 (λ_i 405), ϕ_o = 0.06 (λ_i 436) E_a 14 τ_{-58} 2.1 h, Ref.[30–33)]
62	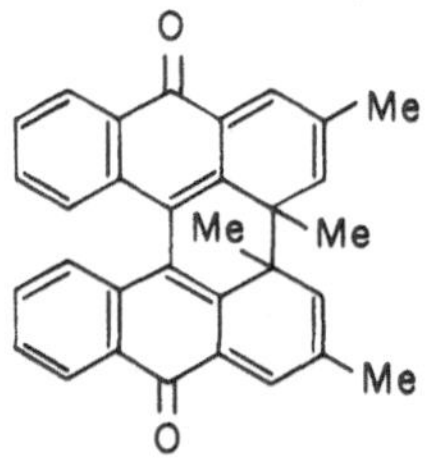 [structure]	C photoisomer, from 1,3,1′, 3′-tetramethyl bianthrone, 5×10^{-5} M, T or MTHF at $-90°$ λ_i 405. λ_{max} I: 600(ϵ 2,600), II: 480, 450, c 30% E_a 14 (MC) τ_{-50} 58 min, ϕ_c 0.6 (λ_i 405), ϕ_o 0.05 (λ_i 436, $-90°$, MTHF) Ref.[30–36)]

Table 9. 4a,4b-Dihydrophenanthrenes with 7 conjugated rings derived from dixanthylidene and its dithio analog

63	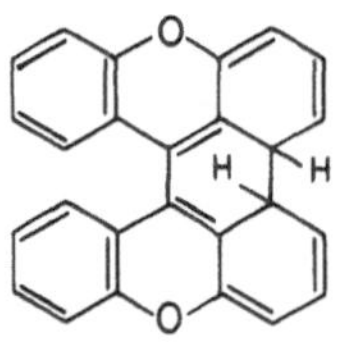 [structure]	Modification C: from dixanthylene λ_i 366 at 0° in MCH/IH c 12% λ_{max} I: 500, II: 410, 390, III: 325 tr -20 to 90°, E_a 11, τ_{55} 13.9 s, Modification P: from dixanthylene, by flash photolysis at +20, in MCH/IH. λ_{max} I: 520, II: 420 tr -23 to 20°, E_a 14.6. τ_{-23} 10.7 s, Ref.[37)]

Table 9 (continued)

64	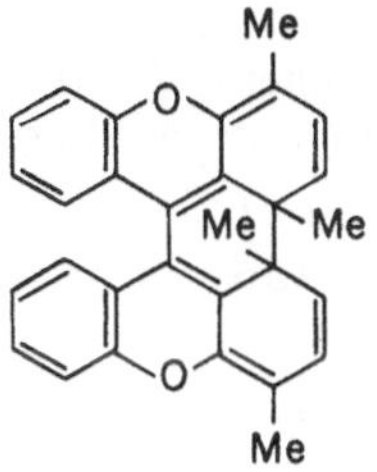 C photoisomer, from 1,4,1′,4′-tetramethyldixanthylene, 3.5×10^{-5} M, MCH/MCP, λ_i 366 at −90°, c 80% λ_{max} 610, 415, ϕ_c (λ_i 436, −80°) 0.2 E_a 12; $\tau_{-70°}$ 4.1 h, τ_{-50} 20.3 min, Ref.[38, 39]
65	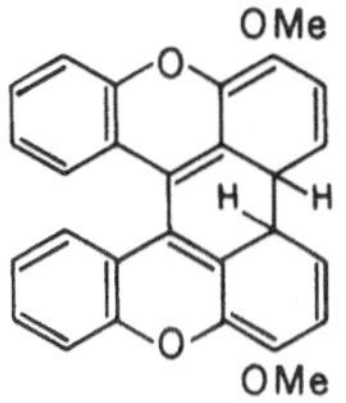 C photoisomer, from 4,4′-dimethoxydixanthylene, 4×10^{-5} M, MCH/IH, λ_i 366 at −30°, λ_{max} I: 540, II: 420, III: 350, Ref.[39]
66	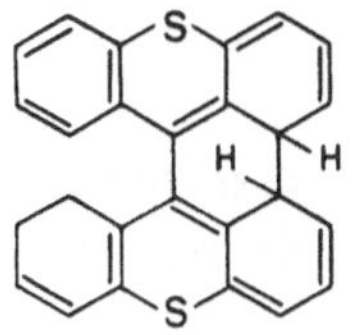 P isomer, from bithioxanthene, 4×10^{-5} M, MCH/IH, by flash photolysis. λ_{max} I: 510, II: 410. c 2%, tr −8 to 80°; E_a 15; $\tau_{0°}$ 0.7 s, light stable. Ref.[40]

Tables 1–9

Abbreviations and Symbols: Me = methyl, Et = ethyl;
Solvents: MCH = Methylcyclohexane; MCP = methylcyclopentane; IH = isohexane; IO = isooctane; H = hexane; MTHF = 2-methyltetrahydrofuran; A = acetonitrile; T = toluene; MC = methylene chloride; E = ethanol.

λ_i = Irradiation wavelength, in nm.
UV^a = Light from medium or high pressure mercury arc filtered through cobalt-nickel sulfate solution; λ_i – 220–330 nm.
UV^b = Near UV light, ca. 350 nm.
λ_{max} = Absorption maximum, in nm. I: first band, II second band, III third band, Sh-shoulder.
ϕ_c = Cyclization quantum yield.
ϕ_o = Ring opening quantum yield.
tr = Temperature range of thermal ring opening measurements.
E_a = Activation energy, Kcal/mole, for thermal ring opening.
τ = Half-life time for thermal ring opening.
E_c = Activation energy of photocyclization, Kcal/mole.
E_o = Activation energy of photochemical ring opening.
Temperature (°C) denoted in parentheses or as subscript.
Ref. = References to literature.
* = Previously unpublished results.
c = Conversion into DHP (%).
D = Optical Density.

a. Systems derived from cis-stilbene and its simple derivatives (*1–22*, Table 1).
b. Systems derived from [2.2] metacyclophanene (*23–26*, Table 2).
c. Three-ring heterocyclic systems (*27–33*, Table 3).
d. Four-ring systems derived from 1-phenyl-2-naphthyl ethylenes (*34–40*, Table 4).
e. Five-ring systems derived from 1,2-dinaphthylethylenes (*41–47*, Table 5).
f. Six-(and more) ring systems derived from 1,2-diarylethylenes (*48–55*, Table 6).
g. Seven-ring systems derived from 10,10′-diH-dianthrylidene (*56–58*, Table 7).
h. Seven-ring systems derived from bianthrone (*59–62*, Table 8).
i. Seven-ring systems derived from dixanthylene (*63–66*, Table 9).

[2.2]metacyclophanene		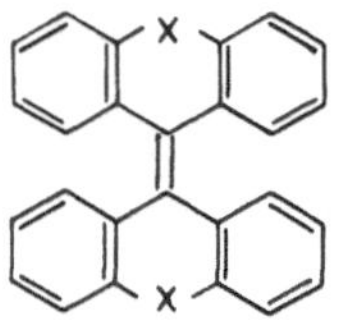 X
	10,10′-diH-dianthrylidene	CH_2
	bianthrone	C=O
	dixanthylene	O

As we shall see later several properties of these molecules such as stability, formation and cleavage quantum yields, and their temperature dependence, can be best treated on the basis of this classification.

III A Molecular Structure

The majority of the 4a,4b-dihydrophenanthrenes listed in Tables 1–9 are definitely thermally[3] unstable at room temperature or below and undergo rapid dehydrogenation by molecular oxygen according to the overall process D′.

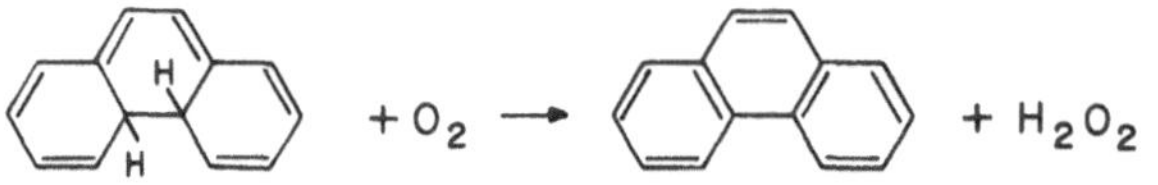

$$+ O_2 \rightarrow \quad + H_2O_2 \qquad D'$$

Thus up to the writing of this review no pure DHP (with the exception of the keto isomer *22*) could be isolated and subjected to usual organic chemical methods of structure determination such as elemental analysis, mass spectroscopy and X-ray crystallography. On the other hand a very wide body of (less direct) structural data[11, 14)] is available which leaves little doubt that the structures of DHP and of

3 e.g. with respect to the dark process C, in contrast to the photochemical ring opening, process B.

its derivatives are as given in Tables 1–9. Some of the more important results helping to establish the molecular structure of these compounds are:

a. Structure of oxidation products[4, 6, 11, 24, 25].
b. Cryoscopic molecular weight determination of *12*[11].
c. Proton NMR spectroscopy of *12*[11, 14], *25*[17], *41*[24], *44*[25, 26], *61*, *62*[30, 31] and of *64*[39].
d. Photochemical ring cleavage of DHP to give back the cis-stilbene molecule[11].
e. Monomolecularity of photocyclization process[11, 30].
f. Electronic spectra[11].

The *trans*-conformation of the 4a and 4b hydrogens (and a C_2 molecular symmetry for *1*) is suggested by several conclusive experimental findings. Among these we should mention:

a. Ozonolysis of *22*, giving dℓ-butane-1,2,3,4-tetracarboxylic acid[41],
b. Absolute asymmetric synthesis of *optically active 44* using both right and left handed circularly polarized light[42],
c. Stepwise transfer of 4a and 4b H atoms to molecular oxygen as required by the trans-conformation (see below)[43–47].

Theoretical analyses of the reaction path of photocyclization point to the same conclusion. Thus the qualitative state correlation procedure clearly indicates that photocyclization takes place by a conrotatory process in the Orbital Symmetry Conservation sense[11], requiring a C_2 molecular symmetry in *1* and in its symmetric congeners. The same conclusion were reached in the subsequent numerical analysis of the photocyclization of *1* and of *44*[48, 49]. The detailed molecular structures of these two molecules and of *61* have been calculated by semi-empirical energy minimization procedures[45, 49, 50] (cf also Ref.[15]).

III B Transient Conformers of 4a,4b-Dihydrophenanthrenes

Nonequilibrium process such as photochemical reactions provide paths leading not only to photoisomers formed in their most stable conformation but also to labile conformations of such photoisomers. A well-known case is that of the labile D conformers of the twisted B photoisomers in bianthrones[30]. The labile D conformers are formed at low temperatures and very high viscosities and revert to the stable B modifications at slightly higher temperatures. Essentially similar effects have been observed with 4a,4b-dihydrophenanthrenes. Thus at room temperature both *44* and *46* are formed by the thermal decay of unstable precursors (Y forms) with half-life times of 10 sec for *44* and 0.3 sec for *46*[24, 25, 27]. The activation energies for the process Y → DHP are of the order of 15 kcal/mole. The visible absorption band in these conformers is slightly red shifted relative to the stable forms, but the intensities of the vibrational components are quite similar. Very recently[24b], an energy minimalization exploration of the potential surface of *44* has revealed the presence of a higher energy conformer derived from the previously suggested geometry[49]. These two conformers of *44* differ significantly in their $H_{14} - H_{15}$ and $H_{4a} - H_{14}$ distance[24b].

A very similar situation has been observed in *6*[62]. Similar but less well studied indications for metastable products have been obtained for the DHP's formed from the m-substituted stilbenes[57]. Some observations about time-dependent changes in the absorption spectrum of *1* could well be due to such processes[7, 24a].

In this context one should also consider the formation of cis-4a,4b-dihydrophenanthrenes (e.g., 4a and 4b hydrogens in cis conformation). While forbidden as an excited state concerted conrotatory process (see Sect. VI D) it could possibly take place by another route. Such cis-conformers would be less stable than the normal trans-conformers[45].

Table 10. ^{1}H Chemical Shifts of 4a,4b-Dihydrophenanthrenes. Atoms are numbered as in Tables 1–9

	Atoms	δ (ppm vs TMS)
12[11,14),a]	4a,4b methyls	1.47
	1, 3, 6, 8 methyls	1.68
	9, 10	5.35
	2, 4, 5, 7	5.60
25[17),b]	4a,4b methyls	1.89 (d, J = 2.7 hz)
	9, 10	4.8 (d, J = 2.7 hz)
	1, 2, 3, 6, 7, 8	5.07 multiplet
	4′, 5′	2.1–2.6
41[25),b]	4a,4b	3.3
	9, 10	6.0
	3, 4, 5, 6	6.4, 6.6
44[25,26),b]	4a,4b	4.0
	9, 10	5.9
	1, 2, 7, 8	6.4
	11–18	7.0
61[30,31),c]	1,1′ methyls	1.74
	2, 3, 4, 2′, 3′, 4′	6.5–6.73
	5–8, 5′–8′	8.18
62[30,31),c]	1,1′ methyls	1.67
	3,3′ methyls	2.05
	2, 4, 2′, 4′	6.4
	5–8, 5′–8′	8.16
64[39),b]	1,1′ methyls	2.8
	4,4′ methyls	2.32
	2, 3, 2′, 3′	6.29
	8,8′	7.35

[a] At 60 MHz [b] at 90 MHz [c] at 100 MHz; d = doublet

IV NMR Spectra

The procedure developed in the initial NMR study of *12*[11, 14] has been used in subsequent studies of *61, 62*[30, 31], of *41* and *44*[25, 26], of *25*[17], and of *64*[39]. The DHP derivative can be obtained only in photostationary concentrations (at most), in reaction mixtures containing both cis- and trans-isomers of the 1,2-diaryl ethylene. Under such conditions the NMR signals due to the nuclei of the DHP derivative are identified as those which disappear following photochemical ring cleavage (process B). This process yields only the cis-isomer of the parent ethylenic compound. In addition to their structural value, NMR studies in this field allow to verify the conversion estimates obtained from optical studies[11, 14]. In *61, 62* and *64* proton NMR proves that photocyclization takes place between 1 and 1′ atoms and not for instance between atoms 1 and 8′ or between atoms 8 and 8′ (numbering as in *61* in Table 8). Table 10 provides a summary of the ^1H chemical shifts of *12, 25, 41, 44, 61, 62,* and *64.* The atoms of the DHP moiety are numbered as in *1* (see Tables 1–9 for details).

V Electronic Spectra

A Absorption Spectra

1 Energies and Intensities

The basic chromophore in 4a,4b-dihydrophenanthrenes is a highly folded fully conjugated hexa-ene bridged by the central two atom unit of carbon atoms 4a and 4b (*1* a). This bridge acts both as a poly alkyl substituent and also as a skeletal constraint.

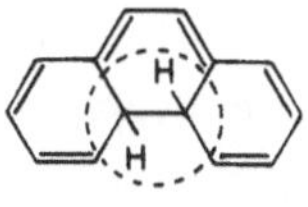

(1a)

This description seems quite adequate for understanding the principal features of adsorption spectra of the 4a,4b-dihydrophenanthrenes as summarized in Tables 1–9, and also suggests several interesting comparisons with the absorption spectra of other polyenic systems[52]. The first (visible) absorption band of 4a,4b-dihydrophenanthrenes is responsible for their intense colours. Two typical spectra, of the three-ring systems *1* and *12* are given in Fig. 1 and 2. These spectra as well as those of the other molecules are composed of three bands denoted I, II and III in order of increasing energy, the first two being the most readily observable. The visible band (I) is broad and usually devoid of vibrational structure excepting those cases listed in Table 13. The extinction coefficients range from ca. 2×10^3 to ca. 2×10^4. The second band is much narrower and shows as a rule several well resolved vibrational components. This and the third band are stronger than the first.

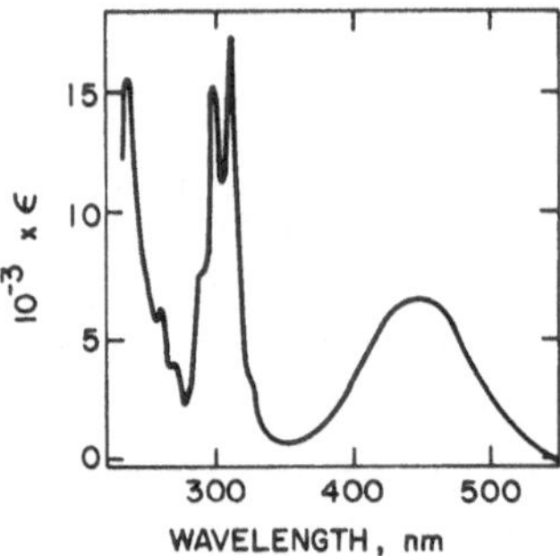

Fig. 1. Absorption spectrum of *1* in MCH/IH at 0 °C[11)]

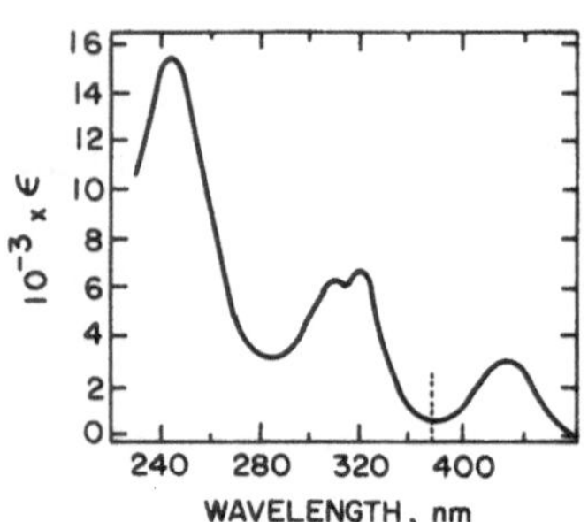

Fig. 2. Absorption spectrum of *12* in MCH/IH at 25 °C[11)]

The simplest theoretical description of the electronic absorption spectra of 4a,4b-dihydrophenanthrenes[11)] seems to be provided by Simpson's exciton theory[53, 54)] of the spectra of polymers. Compared with equally applicable but more complicated MO treatments (e.g. see Ref.[49)] and [55)] for π-electron SCF MO analyses of *1, 44* and of *45*), Simpson's model offers (at least for *1*) some advantages such as numerical simplicity and sufficient transparency without losing too much of physical meaning. In the case of *1* Simpson's exciton model predicts the correct number of transitions and gives estimates of their energies and of their relative intensities.

The exciton model of polyene spectra assumes that each excited state of the polyene may be described by a linear combination of basis states, each having only one (singly) excited ethylenic unit. Only states with neighboring excited ethylenic units can interact.

The basic quantities in this model are E_v the π-π^* excitation energy of ethylene, and Γ, the interaction energy of two basis states with neighboring excited ethylenic-units. Their values are $E_v = 7.60$ ev, $\Gamma = -2.54$ ev.

The formal similarity with the Hückel model is obvious: The expressions for the n double bond polyene in the Simpson model are entirely equivalent to the expressions for the n π-electron system in the Hückel model.

Thus for the case of the linear polyene, the p-th transition energy E_p is given by

$$E_p = E_v + 2\,\Gamma \cos\,[p\,\pi/(n+1)] \qquad (1)$$

and the expression for the coefficient C_{pj} of the j-th excited ethylenic unit in the p-th excited state is

$$C_{pj} = [2/(n+1)]^{1/2} \sin\,[pj\pi/(n+1)] \qquad (2)$$

The transition moment integrals $(I_p)^{1/2}$ can be expressed in terms of the ethylene transition moment M as

$$(I_p)^{1/2} = M\,\Sigma\, a_j C_{pj} \qquad (3)$$

where a_j is the appropriate trigonometric factor for the j-th unit. In the case of *1*, best agreement between calculated and observed transition energies was obtained by lowering E_v slightly to $E_v = 7.20$ ev, to account for both the cis-geometry of double bonds and the effect of the central 4a,4b unit. Equation 1 predicts three transitions (e.g. fundamental transition p = 1 and two overtone bands, p = 2 and 3) with energies below 6.2 ev (200 nm); the calculated energies (Table 11) agree quite well with the observed values.

Table 11. Observed and calculated electronic transition energies of *1*, in ev.[11)]

p	1 (I)	2 (II)	3 (III)
E_p (calculated)	2.62	4.04	5.58
E_p (observed)	2.75	4.08	5.23

The observed order of the three transition intensities, $I_2 > I_3 \gg I_1$ is quite well reproduced by the present model: using Eq. 3, with the coefficients C_{pj} calculated by Eq. 2 (listed in Table 12) we obtain: $I_1 : I_2 : I_3 = 0.21\ M^2 : 1.69\ M^2 : 1.51\ M^2 =$ $= 0.21 : 1.69 : 1.51$.

Table 12. Coefficients C_{pj} of j-th excited unit in p-th excited state of hexaene

p \ j	1	2	3	4	5	6
1	0.232	0.418	0.521	0.521	0.418	0.232
2	0.418	0.521	0.232	−0.232	−0.521	−0.418
3	0.521	0.232	−0.418	−0.418	0.232	0.521

The diagrams of bond transition moments for the three excited states of *1* are given in Fig. 3. The comparison with the bond transition moment diagrams of the all-trans-hexaene (right hand side of Fig. 3) explains readily two intensity features peculiar to *1*:

a. The first transition of *1* is much weaker ($\epsilon \sim 7 \times 10^3$) than the first transition of the all-trans-hexaene e.g. $\epsilon_1 \sim 10^5$ for first transition of $C_6H_5(CH{=}CH)_5CHO$, cf Ref.[52)], Ch. 13).

b. The second and third transitions of *1* are stronger ($\epsilon_2 \sim 2.2 \times 10^4$) than the comparable transitions ("cis bands") in all-trans-$C_6H_5(CH{=}CH)_5CHO$ e.g., $\epsilon_2 \sim 5 \times 10^3$, $\epsilon_3 \sim 10^4$.

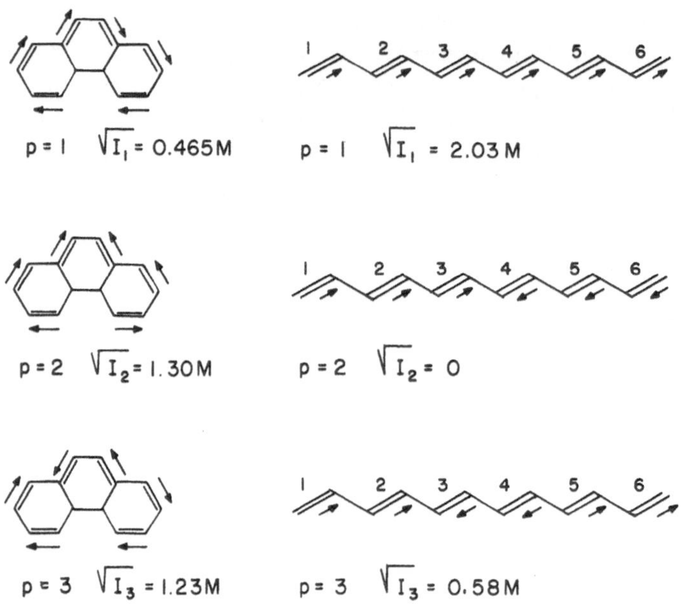

Fig. 3. Bond transition moment diagrams (Exciton model) for first three electronic transitions of 4a,4b-dihydrophenanthrene and of the linear hexaene

2 Effects of Benzo Annelation

Benzo-annelation at one of the double bonds a, c, g, or i (see *1* b) shifts the first maximum to higher energies (e.g., in *1*, λ_{max} = 450 nm. In *34* λ_{max} = 424 nm; in *37*, λ_{max} = 422 nm). Di-benzo-annelation at two of these double bonds has an even more pronounced effect. Thus in *41* (a and i annelation) and in *42* (c and i annelation)

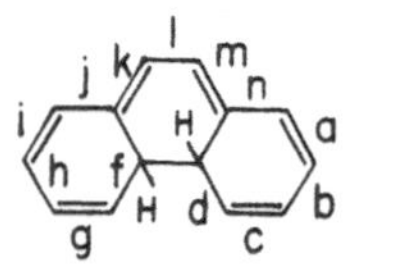

λ_{max} is shifted to 410 nm; in *44* (c and g annelation) it is shifted to 422 nm. The explanation of this effect is quite straight forward. Benzo-annelation across a double bond substitutes a bond with half double bond character for a double bond. In terms of the exciton model this decreases both the length of the interacting system and the strength of the interaction.

Benzo-annelation across a single bond (e.g. b or h) such as in *43* (λ_{max} = 603 nm) or in *45* (λ_{max} = 572 nm) produces bathochromic effects, as a result of the extension of the conjugated system over two additional double bonds. These effects are reproduced quite well by the π-electron M.O. calculation[49].

3 Vibrational Structure

As was mentioned previously, the first absorption band (I) of 4a,4b-dihydrophenanthrene is usually structureless and rather broad, having a half height width ($\Delta X_{1/2}$) of 4700 cm^{-1} in *1*. This band shows vibrational structure (developed to various extents) only in *12, 13, 15, 37, 38, 40, 43–46,* and in *49–53* (see Table 13). The observed vibrational spacings, usually 1200–1400 cm^{-1}, correspond very probably to an excited state-totally symmetric stretching mode (ν_s) of the C–C double bonds such as

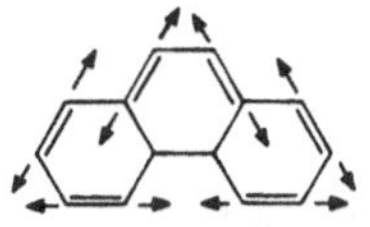

$$\nu_s$$

whose frequency would be lower than the frequency of the same mode in the electronic ground state. This mode is excited by the first electronic transition, as this transition involves a decrease in the bond orders of double bonds, an increase in the bond orders of single bonds and corresponding changes in the bond lengths[51, 56]. A progression of up to four components in this mode can be observed, the *second* component being usually the most intense. The information obtained thus on the vibrational structure of the first transition can be applied also to those molecules in which the vibrational structure of this transition cannot be resolved. This situation is the result of the excitation of lower frequency quanta (probably skeletal out of plane bending modes) in combination with the stretching mode. In sterically hindered molecules which show such resolved spectra, the excitation of these low frequency modes would be less probable because of their steeper potential curve.

Table 13. Excited state vibrational spacings (in cm^{-1}) of 4a,4b-dihydrophenanthrenes[a,b]

1[11]	II: 1412	*46*[24]	I: 1242, 1395
12[11,14]	I: 1125, 1216, 1260	*49*[24]	I: 1260, 1445
	II: 1010		II: 1452, 2073
13[13,*]	I: 1210, 1350	*51*[24]	I: 1687, 2046
15[15*]	I: 1216, 1421	*52*[24]	I: 1324, 1133
20[11]	II: 1426		II: 1386
23[17,18]	II: 1332, 1450	*53*[24]	I: 1204, 1687
24[17,18]	II: 1412, 1546	*56*[28]	II: 1233
25[17,18]	II: 1403, 1421	*57*[28]	II: 1407
37[23–25]	I: 1300, 1200	*58*[28]	II: 1142
38[23]	I: 1200, 900, 1600	*59*[29]	II: 1355
40[23]	I: 800, 900, 1100	*60**	II: 1516
43[24,25]	I: 960, 975	*61*[30–33]	II: 906
44[23–27]	I: 1400, 1500	*62*[30–36]	II: 1389
45[24–27]	I: 1385, 1333	*63*[37]	II: 1251

[a] I and II – first and second excited state, respectively
[b] Spacings listed in increasing order of vibrational levels

The second transition shows in most cases clearly resolved vibrational structure. The mode excited seems to be the same as in the first transition, ν_s, its frequency being usually in the range 1200–1400 cm^{-1}. The geometry changes due to transition II are smaller than those due to transition I, as in transition II the *first* vibrational component is the most intense (and not the second as in transition I[56)]) and the third component in II is much smaller than the first. This vibrational component intensity pattern is one reason why the half height width of transition II is much smaller than that of transition I (e.g., $\Delta X_{1/2} = 2300\ cm^{-1}$ in *1*). Another reason is that transition II seems not to excite combinations with lower frequency modes.

B Emission Spectra

It seems that fluorescence has been only rarely looked for systematically in the 4a,4b-dihydrophenanthrene series. As these compounds undergo usually an efficient excited state ring-opening process (see Sect. VII and also Tables 1–9), only residual (and limited) fluorescence intensity could be expected at most. Thus along this line Naef and Fischer report that *23, 24* and *25* do not fluoresce down to 83 °K[17)]. However, an important exception to these general trends has been observed in the 2-naphthyl derivatives *37, 42, 44,* and *46*[24–26)]. All these molecules are strongly fluorescent at low temperatures, *44* and *46* showing strong fluorescence even at room temperature. As might be expected singlet state emission takes place at the expense of ring opening, the quantum yield of which decreases appropriately. Table 14 summarizes some typical values of fluorescence and ring-opening quantum yields (ϕ_F and ϕ_o, respectively), at −160° and at +20°.

Table 14. Ring opening and fluorescence quantum yields of β-naphthylethylene- derived 4a,4b-dihydrophenanthrenes [24–26)]

	ϕ_o		ϕ_F	
	+20°	−160°	+20°	−160°
37	0.7	0.06	0.0	0.02
42	~1.0	0.002	0.007[a]	0.3
44	0.008	0.0	0.5	0.7
46	0.03	0.0	0.6	0.6

[a] at −20 °C

The efficient emission of molecules in this class is due to the exceptional stability of the lst excited singlet state, as shown by the detailed analysis carried out in the case of *44*[49)], (see also Sect. VII). The fluorescence spectra in this series show the normal "mirror relationship" with the absorption spectra indicating emission from the 1st excited state. The emission vibrational maxima[24–26)] λ and the vibrational spacings ν are given in Table 15.

Table 15. Fluorescence maxima λ, in nm, and vibrational spacings ν in cm^{-1}, of *37*, *42*, *44*, and *46*

37 [a]		*42* [b]		*44* [c]		*46* [d]	
λ	ν	λ	ν	λ	ν	λ	ν
500		461		472		448	
	1430		1355		1384		1408
538 [e]		*492*		*505*		524	
	~1290		~1222		~1620		~1230
579		524		550		560	

[a] at −180 °C; [b] at −160 °C; [c] at −100 °C; [d] at +20 °C; [e] strongest component is underlined

These spectra show uniformly three resolved vibrational components, the second component being always the strongest. The vibrational spacings fall in the range 1600–1200 cm^{-1}, usual values for C–C stretching modes of conjugated polyenes. The Stokes shifts of *44* and *46* are notably small especially when compared to those of the cis-1,2-diarylethylenes[51)].

VI The Photocyclization Process

A Formation of 4a,4b-Dihydrophenanthrenes

As stated previously 4a,4b-dihydrophenanthrenes cannot be isolated pure but can be obtained only in reaction mixtures containing both cis- and trans-isomers of the parent diarylethylene. Thus studies of 4a,4b-dihydrophenanthrenes always depend on the prior development of conditions providing considerable conversion of diarylethylenes into their DHP photoisomers. One requirement for obtaining maximum conversion consists in minimizing the rates of decomposition processes such as photochemical ring cleavage,

$$\text{DHP} \xrightarrow{\Delta} \text{cis-diarylethylene,} \qquad \text{B}'$$

thermal ring cleavage,

$$\text{DHP} \xrightarrow{h\nu} \text{cis-diarylethylene,} \qquad \text{C}'$$

oxidation by molecular oxygen,

$$\text{DHP} + O_2 \longrightarrow \text{phenanthrene derivative,} \qquad \text{D}'$$

and others.

Of these decomposition processes, D′, the oxidation process, can be altogether obviated by using oxidation resistant derivatives such as *12, 58, 60, 61, 62* or *64*. Alternatively 4a,4b-dihydrophenanthrenes can be formed and studied in the complete absence of oxygen e.g., in vacuo or in a nitrogen or argon atmosphere. More recently[57], in studies of *41* and *44*, process D′ could be sufficiently slowed down by using 2,6-ditert.butyl-4-methyl phenol as oxidation inhibitor[11, 43–47] (see also Sect. VIII). The thermal ring opening, process C′, can be in most cases slowed down sufficiently by working at a low enough temperature.

The other requirements for obtaining a maximum extent of DHP formation depend on the properties of photoreversible systems. In the absence of dark processes such as C′ and D′, the composition of a photoreversible system

$$A \underset{h\nu_B}{\overset{h\nu_A}{\rightleftarrows}} B$$

in a photostationary state depends on the position of the photoequilibrium. The concentrations C_A and C_B of A and B at photoequilibrium, $A \rightleftarrows B$, at wavelength λ obey the relationship,

$$C_A^\infty \phi_A \epsilon_A = C_B^\infty \phi_B \epsilon_B \tag{4}$$

where ϕ_A and ϕ_B are the quantum yields for the $A \rightarrow B$ and $B \rightarrow A$ processes, respectively. ϵ_A and ϵ_B are the respective extinction coefficients of A and B at wavelength λ. Eq. 4′,

$$C_B^\infty / C_A^\infty = (\epsilon_A/\epsilon_B)(\phi_A/\phi_B) \tag{4'}$$

assuming a λ-independent ϕ_A/ϕ_B, implies that the highest conversion of A to B is obtained for a maximum ϵ_A/ϵ_B ratio. Some remarkable effects of irradiation wavelength on the conversion of cis-diarylethylenes ($\equiv A$) to 4a,4b-dihydrophenanthrenes ($\equiv B$) due to the ϵ_A/ϵ_B factor are listed in Table 16.

Table 16. Stationary state conversion to DHP (c, %) and extinction coefficient ratio ϵ_A/ϵ_B as function of irradiation wavelength (λ,nm)

	λ, nm	ϵ_A/ϵ_B	c, %
1 [a, 11]	313	0.16	7
	280	3.4	22
12 [b, 11]	313	0.02	~0
	280	1.91	21
20 [c, 11]	313	0.16	17
	280	2.64	67
41 [b, 24, 25]	313	0.66	29
	334	2.5	50

[a] at 0 °C [b] at 25 °C [c] at −20 °C

The importance of choosing an optimum irradiation wavelength is thus obvious. In the case of tricyclic 4a,4b-dihydrophenanthrenes (Table 1) 280 nm gives the highest conversion as this wavelength corresponds to the absorption maximum of the starting cis-diaryl ethylene. Irradiation at a λ where only *B* absorbs ($\epsilon_A = 0$) leads of course to complete decomposition of *B*, while irradiation where only *A* absorbs ($\epsilon_B = 0$) results in complete conversion to *B*. In the case of 4a,4b-dihydrophenanthrenes in the visible, we have $\epsilon_A = 0$ as only the DHP absorb. This last factor requires the exclusion of visible light (and also of shorter inactive wavelengths, e.g. $\lambda > 330$ nm, for the case of *1–33*) when maximal conversion to DHP is desired.

B Kinetic and Mechanistic Studies of the Photocyclization

In addition to the definite conclusions of the theoretical studies (see Sect. VI C) which rule out a "hot ground state" process there are numerous experimental findings which clearly suggest that 4a,4b-dihydrophenanthrene-like photocyclizations (with few exceptions, see below) take place in the first excited singlet state of diarylethylenes[11, 48, 49, 58]. Among those findings we should note the complementary relationship of cyclization and cis-fluorescence quantum yields[11, 24, 25], and the lack of sensitization of the cyclization process by triplet sensitizers and of quenching of same process by triplet energy quenchers[11, 58, 59].

Excited state process of cis-diarylethylenes and their temperature dependence

The first excited singlet state (S_1) of cis-diarylethylenes undergoes the following processes (see Fig. 4): fluorescence (F)[60] internal conversion (IC)[60], photocyclization (PC) and intersystem crossing to the triplet states (ISC)[61] with ensuing isomerization to the trans-isomer (ISO)[61]. In cis-diarylethylenes the intersystem crossing process is neither activated (temperature dependent) nor viscosity dependent[60]. The other two processes which compete with F in the deactivation of S_1 into a second intermediate (analogous to ^{3}*67*B) are IC and PC. IC is strongly viscosity dependent while PC shows only very limited viscosity dependence[13, 60]. The strong viscosity dependence of IC[60] usually overshadows any possible temperature dependence excepting 1,2-diphenylcyclopentene, in which molecule clearcut dependence of IC on temperature is evident. PC is usually temperature dependent, the only exceptions being the precursors of *23, 24* and *25,* which are also completely nonfluorescent[17]. The observed temperature dependence of PC corresponds to empirical activation energies E_c (see Table 17) within the range of 1–5 Kcal/mole, with the exception of *44.* The observed temperature dependence of fluorescence of

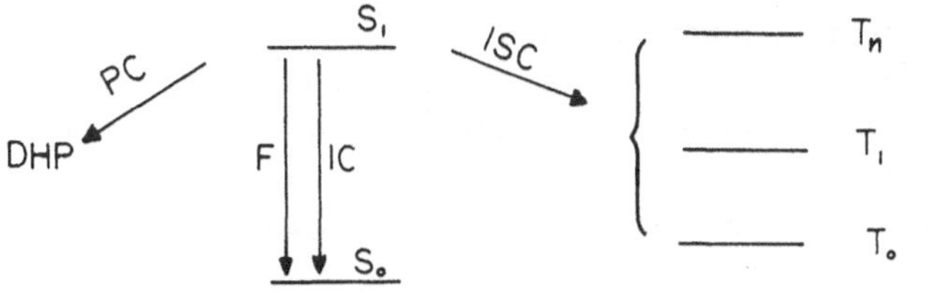

Fig. 4. Jablonski-type diagram of cis-diarylethylene excited state processes

Table 17. Empirical activation energies (Kcal/mole of photocyclization (E_c) and of fluorescence (E_F) of cis-diarylethylene precursor and of ring opening of DHP (E_o)

	1[a]	*20*[a]	*34*[b]	*37*[b]	*41*[b]	*42*[b]	*44*[b]
E_c(cis)	1.2	2.5	5.5	4.5	2.5	2.5	10.0
E_F (cis)	–	2.5	3.5	3.0	3.5	4.0	4.5
E_o (DHP)	0	0	0	3.0	0	4.0	6.5

[a] Ref. [11)] [b] Refs. [24, 25)]

the cis-diarylethylenes of Table 17 (e.g., *20–44*) seems largely due[60)] to that of the internal conversion. In the majority of cases (e.g., excepting *44*) E_F and E_c are of similar magnitude, a finding which suggests a common activation mechanism for both IC and PC. This possibility is supported by the lack of fluorescence in *23–25* and the temperature independent photocyclization in these systems. Such effects of a bridge at the 3,3′positions of cis-stilbene as observed in *23–25* show that the activation energies of IC and of PC are not required for obtaining additional electronic promotion to a higher excited electronic level. Instead, as suggested earlier[11, 48, 51)], these results indicate that vibrational modes involving stretching and torsion of the α–α', 1–α and 1′–α' bonds are involved in such processes. In the case of PC (with the exception of *23–25*) activation energy would be also required to overcome considerable "closed shell repulsion". In both cases of IC and PC, radiationless transition theories (for discussions see Refs. [28] and [51)], predict dependence of rate on square of vibrational overlap integrals between initial and final (including virtual) states. Between widely separated states largest values of these integrals are expected for those vibrations (modes) which involve large changes in their leading internal coordinates[51, 56)]. In the case of cis-diarylethylenes[51)] these are modes involving the central 1–α, α–α' and 1′–α' bonds.

Triplet State Photocyclizations

While the photocyclization of most cis-diarylethylenes takes place from the 1st excited singlet state, there are two related systems which probably undergo cyclization from the first triplet state.

N-Methyldiphenylamine (*29* A) undergoes photocyclization to give the 4a,4b-dihydrophenanthrene-like product *29*[21)]. The rate of the decay of the absorption of the triplet of *29* A at 540 nm corresponds to the rate of growing-in of the absorption of *29* at 610 nm which proves that the cyclization takes place from ^{3}*29* A[21)]. Pentahelicene, *67*, (obtained by loss of H from *44*) gives *67* A by photocyclization-dehydrogenation. In this case no dihydro intermediate (*67*B) can be isolated but flash photolysis experiments seem to indicate that the triplet of *67* (^{3}*67*, absorption maximum at 515 nm) undergoes cyclization to give another intermediate (absorption maximum at 430 nm), presumably ^{3}*67*B which would then dehydrogenate to give *67* A[24, 62)]. In the benzopentahelicene *68*, which does not undergo cyclization[63)] only ^{3}*68* could be observed. This transient was not converted into a second intermediate (analogous to ^{3}*67*B).

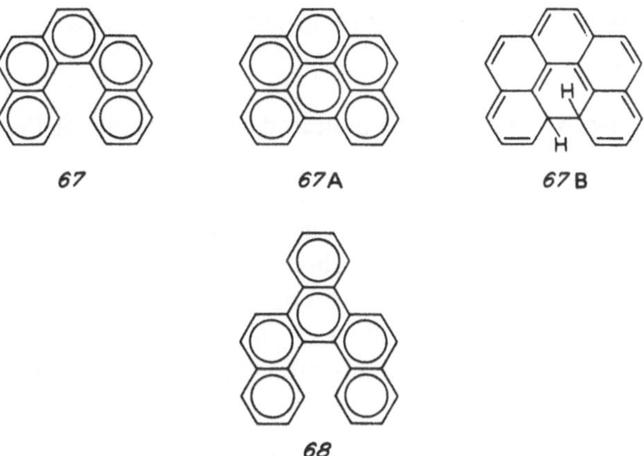

C Structural Effects on Photocyclization Reactivity

In addition to the extensive information on the excited state cyclization reactivity summarized in Tables 1–9, a wide body of reactivity data is available from studies of the direct photoaromatization process (D), the rate limiting step in such cases being the photocyclization process itself. This wealth of systematic and rather detailed excited state reactivity data is undoubtedly unique, providing numerous opportunities for theoretical studies of excited state reactivity some of which will be reviewed in Sect. VI. D.

Photocyclization rate constants are the primary and most direct reactivity measures. However when such data are unavailable, photocyclization quantum yields (ϕ_c in Tables 1–9) can serve as reactivity measures provided closely similar or parallel processes are considered. Under such conditions photoequilibrium concentrations (c in Tables 1–9) or even chemical yields (for the direct photoaromatization) are equally useful. Photocyclization reactivity can be expressed either relative to that of the parent molecule or for a series of parallel processes,

$$A \begin{cases} \rightarrow B \\ \rightarrow C \\ \rightarrow D \end{cases}$$

as relative to the reactivity of one path, say $A \longrightarrow B$.

A variety of distinct molecular perturbations can modify strongly the photocyclization reactivity of 1,2-diarylethylenes. The following classification of such perturbations suggests the main factors responsible for the observed effects but considering the complexity of the problem should not be accepted at times without due caution.

a Electronic Effects

1 Substituent Effects

para Substitution. The reactivity is very markedly depressed by strong electron donating groups, at the 4 position, e.g., amino, dimethylamino and methoxy and by strong electron attracting groups such as cyano, nitro, acetyl or benzoyl[11,15,48,73].

Other substituents, such as alkyl, fluoro, chloro and bromo show much weaker effects.

ortho Substitution. Decreased reactivity seems to be indicated by the results for the chloro group[74].

meta Substitution. This type of substitution can give rise in principle to both 2 and 4 substituted DHP's. These two cyclization paths were observed for methoxy[15], amino[15], and trifluoro methyl[74] substitution. Amino and methoxy substituents at the meta position produce a remarkable reactivity enhancement[73]. The methyl group shows a lesser effect, the halogens are without any effect and electron attracting groups have a strong deactivating action.

α,α'-Substitution. Fluoro substitution results in some deactivation[11], while the cyano (Table 1) and carboxylate[74] seem to have only moderate effects.

2 Hetero-atom Effects

Skeletal heteroatom substitution on the aryl groups and hetero atom ring annelation have been especially well studied in the present context. Thus in addition to the examples listed in Tables 2, 3 and 4, much data has been obtained in studies of the direct photoaromatization. As a whole, the photocyclization process viewed as a hexatriene-cyclohexadiene ring closure is only moderately influenced by hetero atom substitution as can be judged by comparing results for *1* (Table 1) with results for *26* (Table 2), and for *27* and *33* (Table 3). Branching into parallel cyclization paths has been studied experimentally and theoretically (see Section VI·D and Ref.[20, 75 and 76]), and is to be expected whenever the cis-1,2-diarylethylene can exist in several distinct *s* conformers. In such case the preferred cyclic product is derived from that hexatriene system in which its 1–2 and 5–6 bonds possess the highest double bond character, though other factors (e.g., N–N electrostatic repulsion) play a role, too.

Table 18 lists several aza and diaza diaryl ethylenes in which the photocyclization takes place. The photocyclization of molecules *76–82* and of *86–88* were studied by R. H. Martin et al.[77]. For the literature on the other aza ethylenes see Ref.[20]. Several thia and oxa 1,2-diarylethylenes in which the photocyclization has been reported are listed in Table 19 (see Ref. 76 for original literature). In both aza and thia series parallel cyclization paths were observed in *75*a and *75*b, in *77*a, and *77*b, in *82*a, *82*b, and *82*c, in *83*a, *83*b and *83*c, in *84*a, and *84*b, and probably in *95*a and *95b*.

3 Electrostatic Repulsion Effects

Some s-conformers of cis-1,2-diarylethylenes with aza substitution on both aryl groups show attenuation of their photocyclization reactivity[20, 77]. These conformers

Table 18[a]

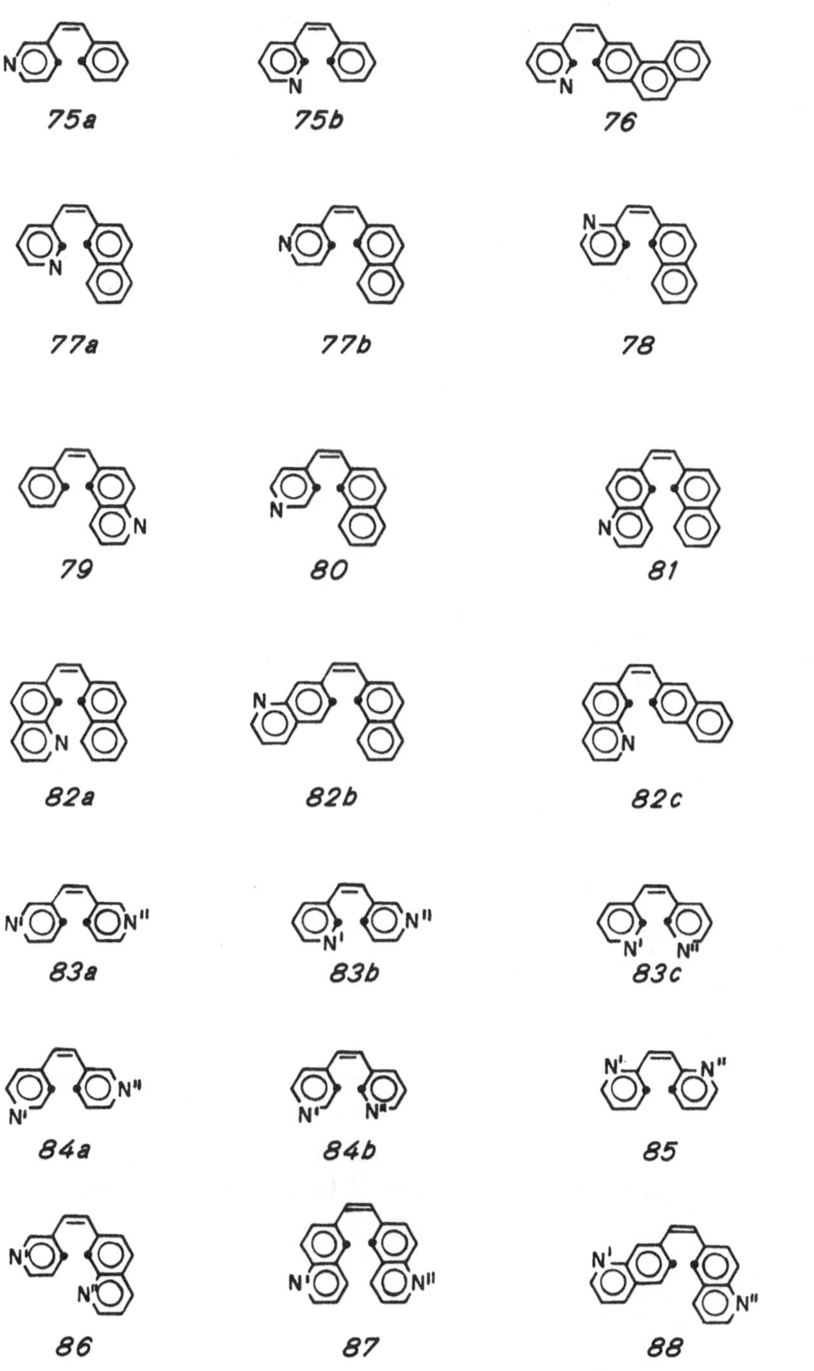

[a] The dotted positions denote the atom pair forming the new bond.

Table 19

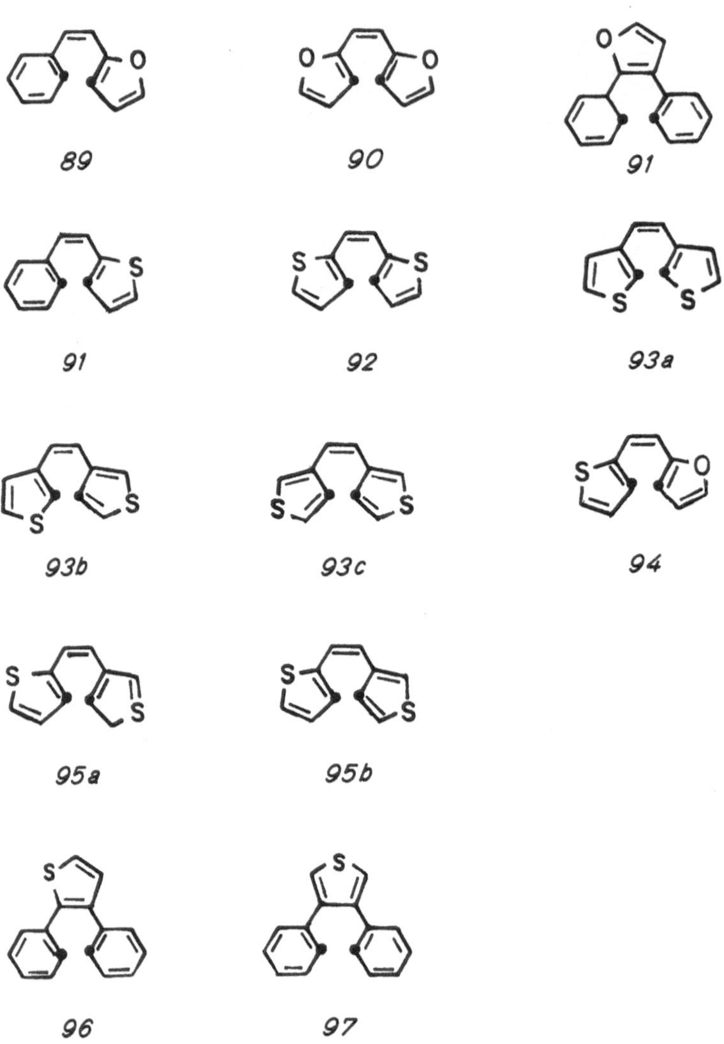

have relatively short N–N distances. Thus *84*b is less reactive than *84*a, and the cyclizations of *86*a and of *88*a do not take place at all[77]. Other systems showing a similar behavior are known[20, 77].

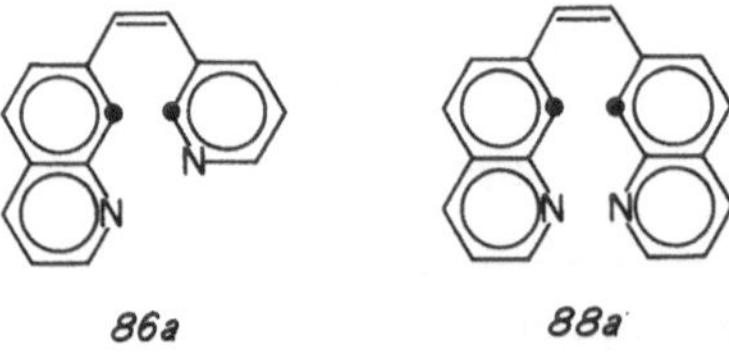

In such systems the theoretical reactivity studies predict similar reactivity for pairs such as *86*–*86*a and *88*–*88*a and the lack of reactivity of *86*a and of *88*a is a consequence of their short N–N distances. These equilibrium values are 5.9 vs 4.4 Å in *86* and *86*a, respectively, and 6.0 vs 4.8 Å in *88* and *88*a[20]. The net charges on the two nitrogen atoms in both ground and first excited singlet states are sizable and roughly equal, about –0.5 e[20]. Thus at the short approach distances required for the formation of the new bond (and even in the initial geometry at the moment of the optical excitation) the unfavored conformers will experience considerably larger electrostatic repulsions than the other conformers. These repulsive interactions act just along the reaction coordinate. In some diaza diarylethylene systems these interactions are large enough to prevent the cyclization altogether[20]. In the case of *dicyclic* and larger aryl groups the critical N–N equilibrium distance is about 5.1 Å. Below this value no cyclization is observed[20].

4 Topological Effects

Annelation of additional aromatic units to the basic cis-1,2-diphenylethylene system exerts strong effects on its inherent reactivity. In the usual MO description these effects can be traced to the effect of the structure of the new skeleton on the highest occupied and lowest unoccupied orbitals at the atoms forming the new bond and therefore can be properly considered as topological effects. As such effects are quite numerous we shall limit ourselvs to only a few examples. Thus o-terphenyl (*103*) does not give any DHP under usual conditions[57].

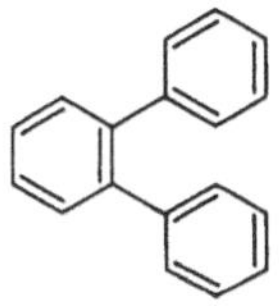

103 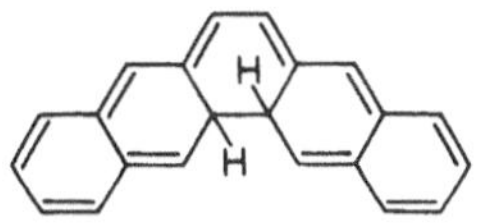 *104* 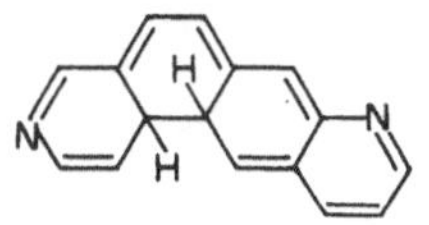 *105*

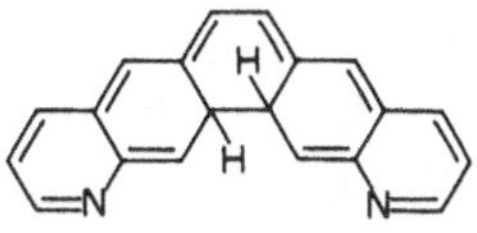

106 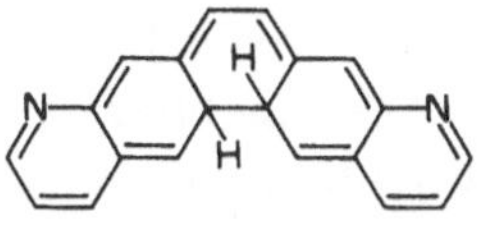 *107*

The same inertness holds for many inoperative cyclization paths that would have given highly quinoid systems such as *104*[49, 76] (cf *44* and *45*), *105*[20] (cf *86*). *106* (cf *87*) and *107* (cf *88*). Under certain circumstances such topological factors can not only modify the AO coefficients at the reacting atoms but can also result in an interchange of the usual topmost occupied MO and lowest unoccupied MO with other orbitals which prevent the photocyclization altogether. This situation has been deduced for the pentahelicenes series (Table 20)[63]. In this series, (Table 20) benzo

Table 20

67	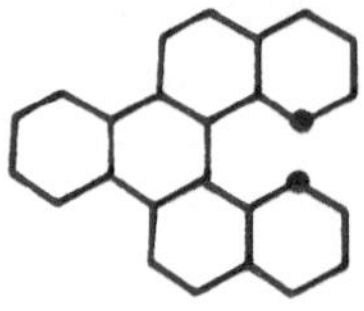	*68*
98	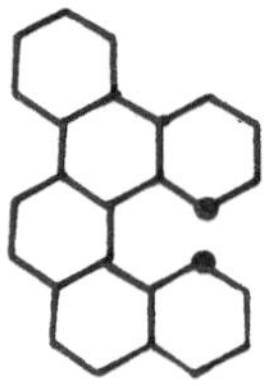	*99*
100	*101*	*102*

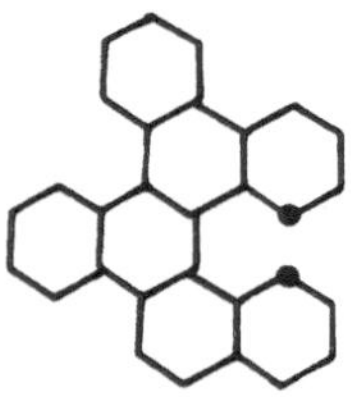

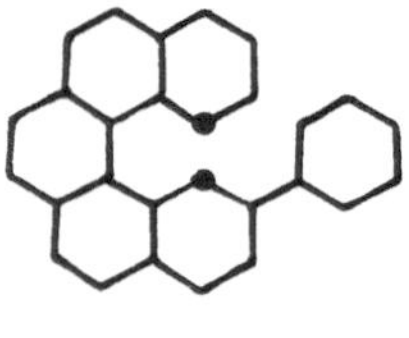

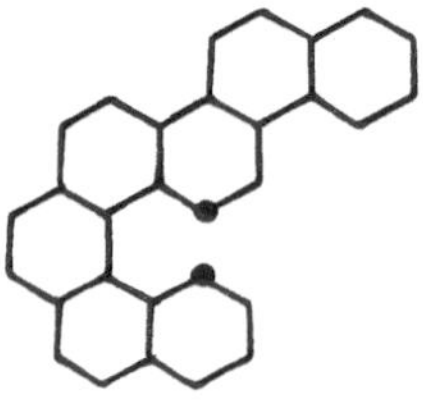

annelation across the 7–8 bond of pentahelicene (*67*) as in *68* results in loss of photocyclization reactivity. The same effect is observed in the pair *98* and *99* (which does not photocyclize). Phenyl substitution (*100*) or other types of benzoannelation (as in *98, 101,* or *102*) do not lead to loss of photochemical reactivity.

b Steric Effects[13, 17–19, 49)]

The influence of steric effects on the photocyclization process is clearly discernible for the ortho and meta substituted stilbenes. In these cases steric repulsion intervenes as the two C atoms forming the new bond approach. This steric repulsion opposes the stabilizing interactions which promote the cyclization process. As a result the reactivity is decreased. Other consequences of such superimposed steric repulsions are either decreased or increased thermal stabilities which will be considered in Sect. VII. In addition to effects at the cyclization region, steric effects in distant parts of a photocyclizing molecule can control the outcome of the reaction. Such is the situation for the 8 and 8′ substituent in the 1,1′ dimethyl dianthrylidene series (cf Tables 7–9).

Steric effects of substituents at the ring ortho position are most evident when comparing the formation of *1* and *12*. The quantum yield for formation of *1*, $\phi_c \sim 0.1$ is decidely larger than for the formation of *12*, $\phi_c \sim 0.04$. In this case the repulsion of the two neighboring methyl groups (4a and 4b in the product) comes into play in the initial part of the reaction coordinate.

Steric effects of groups occupying the meta position of the parent system can be seen in isomeric pairs of the higher systems such as *44–45*, *51–52*, and *53–54*. Thus along the cyclization path leading to *44* closed shell repulsion of H_{4a}–H_{15} and H_{14}–H_{15} are larger than along the path leading to *45* and correspondingly $\Phi_c = 0.02$ for *44* but for *45*, $\Phi_c = 0.06$. The same effects can be recognized in the two other pairs: for *51*, $\Phi_c = 0.03$, compared to $\Phi_c = 0.06$ for *52*; and $\phi_c = 0.004$ for the more sterically hindered *53 vs.* $\phi_c = 0.02$ for the less hindered *54*.

The formation of the sterically hindered *14* shows evidence for deep effects of the nonbonded repulsions of the 4 and 5 methyl groups: This DHP which is extremely unstable at ambient temperatures is nevertheless formed at low temperatures at which the formation of *1* does not take place. In this case the steric interactions in the educt obviously exert a strong influence on a temperature dependent process.

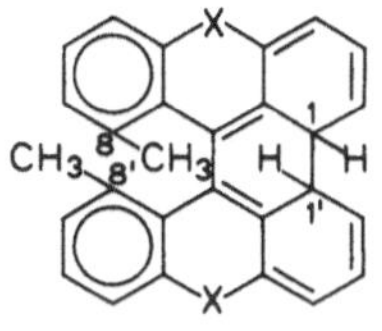

108a 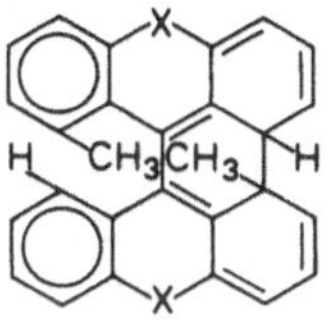 108b 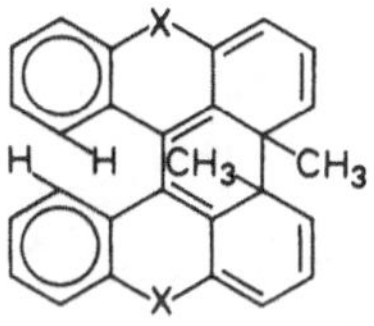 108c

In the 1,1′ dimethyldianthrylidene series X=CH_2, CHOH, C=O and O (*58*, *61*, *62*, and *64*) steric repulsion of the substituents at the 8 and 8′ positions in the final stages of the bond formation at 1–1′ seems to control the outcome of the reaction. This repulsion would be larger for the path leading to *108*a and to *108*b than to *108*c. This is borne out by the calculated strain energy, which for *108*a is higher by 27 kcal/mole than for *108*c[78]. Experimentally only 1,1′-dimethyl cyclization products of type *108*c were observed[28, 30–33, 36–39].

D The MO Analysis of Reactivity in 1,2-Diarylethylene – 4a,4b-Dihydrophenanthrene Systems

The excited state cyclization (process A) and the related excited state or ground state ring opening (processes B and C, respectively) have been the subject of detailed quantitative MO studies[15, 20, 48, 49, 63, 75, 76]. Some insight into the features of the

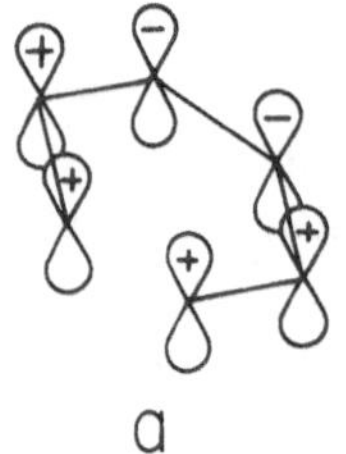

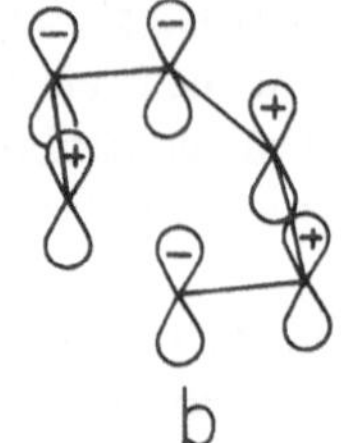

Fig. 5 a and b. Schematic representation of: **a** highest occupied MO, **b** lowest unoccupied MO in hexatriene

ground and excited state reactivity can be deduced by the qualitative approach of Woodward and Hoffmann[80]. Provided that the highest occupied and lowest unoccupied orbitals of educt and product resemble those of hexatriene and of cyclodexadiene one can arrive at the following conclusions for concerted *conrotatory* processes such as A, B and C:

a. The first excited state processes A and B are allowed. b. The ground state process C is forbidden. The simplest way to derive these results is by considering the interactions between the AO'S centered on the 1 and 6 carbon atoms of hexatriene[80)]

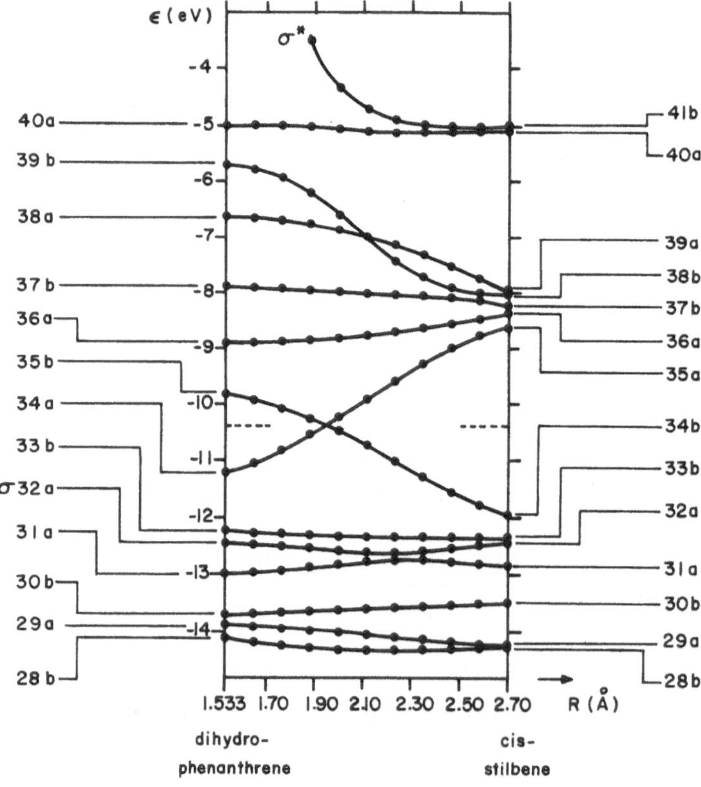

Fig. 6. Orbital correlation diagram for the DHP-cis-stilbene conrotatory path. R is the C(4a) – C(4 b) separation. The dotted line indicates the ground state occupancy limit. The molecular orbitals were computed by the Extended Hückel method[82)]

in the topmost occupied MO (process C, cf Fig. 5a) and in the lowest unoccupied MO (process A and B, cf Fig. 5b). For the more rigorous deviation, see Ref. [81]. The terms "allowed" and "forbidden" refer to the absence or presence of a potential barrier of electronic origin along the reaction path. The crossing of orbitals of different symmetry along the path of a forbidden process gives rise to a potential barrier. In an allowed process as the corresponding orbitals correlate no potential barrier is to be expected on such grounds. For this reason the latter processes are usually strongly preferred. Forbidden processes, on the other hand, can take place only in the absence of competition by other preferred processes and provided they are not rendered impossible by thermodynamic factors. These considerations apply wholly in the DHP-cis-diarylethylene systems, as was shown by the explicit calculations for the DHP-cis-stilbene case[48]. The calculated orbital and state correlation diagrams for the conrotatory paths are given in Figs. 6 and 7. The topmost occupied MO in cis-stilbene, $34b$, is seen (Fig. 6) to go over to the lowest unoccupied MO in DHP, $35b$, while the topmost occupied MO in DHP, $34a$, goes over to MO $35a$ in cis-stilbene. Thus the ground state interconversions involve crossing from a configuration $\ldots\ 32a^2\ 33b^2\ 34a^2$ in DHP into the configuration $\ldots\ 32a^2\ 33b^2\ 34b^2$ in cis-stilbene and are thus forbidden. On the other hand, the excited state paths, configurations $\ldots\ 32a^2\ 33b^2\ 34b^1\ 35a^1$ in DHP, $\ldots\ 32a^2\ 33b^2\ 34a^1\ 35b^1$ in cis-stilbene, have no maxima in the transition state region ($R \approx 1.9$ Å) and are thus allowed. These considerations are made clearer by examining the state correlation diagrams of Fig. 7 for the ground and 1st excited states (A and B, respectively). The potential curve for state B (Fig. 7) exhibits a minimum at $R \approx 1.9$ Å. Thus the excited state processes, from either cis-stilbene or from DHP proceed along this path.

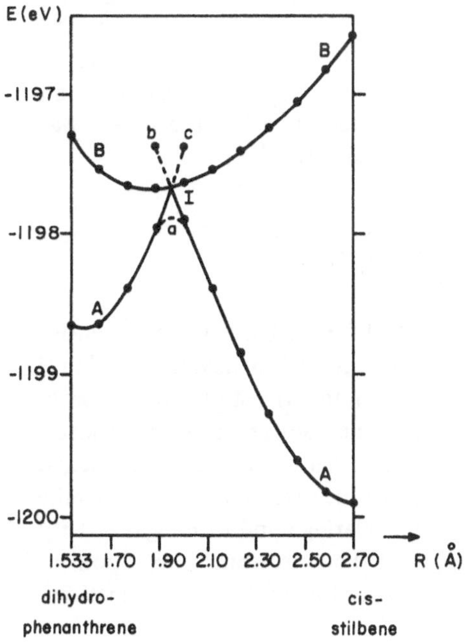

Fig. 7. State correlation diagram for the DHP-cis-stilbene conrotatory path. R is the reaction coordinate, as in Fig. 6. A and B are the state symmetry species. The dotted parts of *b* and *c* are the potential curves for doubly excited cnfigurations. *a* describes the effect of configuration interaction

Crossing with branching into the ground states of product and educt takes place at this point which corresponds effectively to a transition state. The consequences of this interpretation will be examined in a later section.

The ground state potential curve (state A, Fig. 7) shows a potential barrier in the vicinity of the excited state minimum. The dotted parts of branches b and c beyond point I correspond to the potential curves of the doubly excited configurations. Going from DHP to I requires an activation energy of 23 kcal/mole which is lowered to ca. 18 kcal/mole if the C.I. depression of state A is taken into account[48]. Going from cis-stilbene to I along the thermal path is very strongly endothermic ≈ 53 kcal/mole. The potential curve for state A thus illustrates two cases of forbidden process: The thermal conrotatory ring closure which is a clearly unlikely path and the thermal conrotatory ring opening which as such is forbidden, but because of the high energy of DHP vs cis-stilbene nevertheless can take place. The calculated energy difference between DHP and cis-stilbene amounts to 29 kcal/mole which is close to the thermochemical estimates[6b].

The MO analysis of reactivity in processes A, B and C by the explicit computation of energy profiles for the reaction path as described above for the parent system could be undoubtedly carried out also for the other members of this series. However, considering the large size of these systems and the relative lack of perfection of the computations still practical in such cases there are important advantages in resorting to approximate reactivity analyses which depend on the application of Perturbation Theory.

Free Valence Method

Early studies of photocyclization reactivity using Coulson's Free Valence Numbers[83] (F_r) were carried out by Scholz, Dietz and Mühlstadt[84], and by Laarhoven et al.[85]. Relatively good reactivity predictions for several parallel cyclization paths were obtained taking the sums of excited state Free Valence numbers for the reacting C atom pair (ΣF_r^*) as reactivity measure[85]. In this sense a threshold value of $\Sigma F_r^* = 1$ was assumed, $\Sigma F_r^* < 1$ implying lack of reactivity[85] (cf also Ref. [63]).

Electronic Overlap Population Method[15,20,48,50,63,75,76]

Most of the reactivity problems described in Sect. VI C have been treated recently by methods based on the electronic Overlap Population concept as introduced in the Electronic Population analysis[86] of R. S. Mulliken. In fact the studies of the problems of Sect. VI C have established the general usefulness and applicability of electronic overlap populations as reactivity measures for other excited- and ground-state reactions. The most important advantage of the electronic overlap population method of reactivity analysis is the possibility provided for relating bond forming reactivity of two initially nonbonded atoms to the strength of their electronic interaction.

The definition of electronic overlap population, at three different summation levels, is as follows[86].

For a given pair of AO's r and s centered on two atoms, k and l, respectively, the *partial overlap population* $n(i; r_k s_\ell)$ due to the *i*th MO (occupied by N_i electrons) is defined as

$$n(i; r_k, s_\ell) = 2\, N_i C_{ir_k} C_{is_\ell} S_{r_k s_\ell} \tag{5}$$

where as usual C_{ir_k} denotes the AO coefficient and $S_{r_k s_\ell}$ is the overlap integral. This is the primary quantity. Summations over all occupied MO's i gives a *subtotal overlap population* $n(r_k s_\ell)$.

$$n(r_k, s_\ell) = \Sigma_i\, n(i; r_k s_\ell). \tag{6}$$

Further summation over all AO's centered on k and ℓ gives a total overlap population, $n(k, \ell)$

$$n(k, \ell) = \sum_r \sum_s n(r_k, s_\ell). \tag{7}$$

The approximate interaction energy $\Omega_i(r_k s_\ell)$ of an atom pair k ℓ due to the overlap of AO pair (r, s) in MO i is given by the expression[86)]

$$\Omega_i(r_k, s_\ell) = A_\tau I_{rs}\, n(i; r_k s_\ell) \tag{8}$$

The energy $\Omega_i(r_k, s_\ell)$ (overlap energy) is proportional to the partial electronic overlap population due to the *i*th MO and to the average ionization potential $\overline{I}_{rs}$.

For an electron in a carbon 2 P_z AO, $\overline{I}_{rs} \approx -10$ e.v.

A_τ is an empirically determined constant which assumes different values for σ or for π overlap. A_π is considerably larger than A_σ and is roughly equal to 1. Quite parallel to (6) and (7), we have also

$$\Omega(r_k, s_\ell) = A_\tau I_{rs}\, n(r_k, s_\ell) \tag{9}$$

which can be further summed over the AO's to give $\Omega(k, \ell)$. This overlap energy can be written as a sum of σ and of π electron contributions,

$$\Omega(k, \ell) = \sum_{\sigma \text{ pairs}} \Omega(r_k, s_\ell) + \sum_{\pi \text{ pairs}} \Omega(r_k, s_\ell) \tag{10}$$

Thus (8), (9) and (10) imply that electronic overlap populations (partial, subtotal or total) may serve as direct measures of electronic interaction between two atoms k and ℓ. Considering a pair of nonbonded atoms, a positive electronic overlap population $n(k, \ell)$ or a positive change $\Delta n(k, \ell)$ due to electronic excitation or to a change in a reaction coordinate correspond to a stabilizing interaction which favors the bond formation process. Negative values of $n(k, \ell)$ or negative changes $\Delta n(k, \ell)$ correspond on the other hand to destabilizing (repulsive) interactions which oppose bond formation while small values, close to zero, indicate lack of reactivity. For reactive systems the magnitude of the bonding interactions are quite substantial. Thus in the initial reaction stage in hexatriene-like systems (at C–C distances of 2.8–2.6 Å) the excited state electronic overlap populations, $n^*(k, \ell)$ fall within the range of 0.01–0.03, corresponding to interaction energies of 2–10 kcal/mole. In addition to the photocyclization of 1,2-diaryl ethylenes[75,76], and of the thia, oxa and aza analogs[20,76] electronic overlap population analyses were applied also to several other reactivity problems such as photocyclizations of o-quinodimethane and of 1,4-diarylbutadienes[76], photodimerization of 2-methoxy naphthalene, anthracene, trans-stilbene, sorbic acid[76], and acenaphthylene[87]; bond cleavage reactions, additions to the butadiene system and ground state decarboxylations[88].

As an illustration of a typical applications of electronic overlap population for analyzing photocyclization reactivity, Table 21 summarizes the results obtained for the pentahelicenes[63]. n^* denotes the excited state electronic overlap population for the atom pair forming the new bond and Δn is the corresponding change due to the one electron Extended Hückel[82, 86] MO's. For the photocyclizations of 1,2-difuryl ethylenes very similar results were obtained also from minimal basis set *ab-initio* 2.5 Å[63]. In the case of pentahelicenes the benzoannelation in *68* and *98* prevents the photocyclization. For both molecules indeed n^* and Δn are negative indicating destabilizing interaction. For the other molecules, however, both n^* and Δn are positive (and large) indicating significant stabilizing interaction. In the pentahelicene systems, the excited state free valence sum ΣF^* is inapplicable as the values calculated (Table 21) are for all molecules well below the threshold of unity.

The electronic overlap populations in all three cases were calculated from the one electron Extended Hückel[82, 86] MO's. For the photocyclizations of 1,2-difuryl ethylenes very similar results were obtained also from minimal basis set *ab-initio* wavefunctions[76]. The possibility of obtaining useful reactivity analyses from wavefunctions which are easily available even for large systems could prove to be an important practical consideration for further applications of this method. The dependence on $S_{r_k s_\ell}$ in (5) ensures that electronic overlap populations show the desirable physical characteristics for their use as reactivity measures: strong falling-off with increasing interatomic distance and proper directional dependence. This last point is of particular significance for bond formation in polyenes. Thus for two C 2 p_z atomic

Table 21. Comparison of reactivity parameters n^*, Δn and ΣF^* for photocyclizations of pentahelicenes[63]

Path	n^*	Δn	ΣF^*	Observed reactivity
67	0.0599	0.0583	0.984	+
68	−0.0091	−0.0159	0.967	−
99	0.0530	0.0519	0.911	+
98	−0.0050	−0.0078	0.901	−
100	0.0470	0.0524	0.935	+
101	0.0524	0.0518	0.899	+
102	0.0368	0.0328	0.988	+

orbitals with equal phases as for the hexatriene 1 and 6 atoms in Fig. 5a S is positive in a planar system (2 $p\pi - 2p\pi$ overlap). However, in a nonplanar system S would be negative were the local Z axes collinear (2 $p\sigma - 2p\sigma$ overlap). Electronic overlap populations are especially useful for analyzing the reactivity of systems devoid of any symmetry element suitable for the application of the Orbital Symmetry Conservation rules. Another important application of electronic overlap population is the case of parallel reaction paths such as those leading to *42* and *43*, to *44* and *45* and many others[76].

VII Ring Opening Processes

Both thermal and photochemical processes yield exclusively the cis-isomer of the 1,2-diaryl ethylene[11]. As mentioned in Section VI D these two processes proceed through conrotatory paths. The excited state ring opening is an allowed process while the ground state ring opening is forbidden but nevertheless proceeds readily due to the relatively high energy of the ground state of DHP. As a forbidden process, the thermal ring opening requires crossing over a potential barrier of electronic origin. The experimentally determined activation energies for this process (E_a in Tables 1–9) are usually close to ca. 15 kcal/mole. The influence of steric factors on these activation energies is well documented. High values (E_a = 29 kcal/mole in *46*, 23 kcal/mole in *44*) are observed whenever symmetrical nonbonded repulsions maintain the DHP molecule in a 'potential box'[49]. Low values are obtained whenever steric repulsion in the direction of ring opening are not opposed by repulsions acting in the opposite sense. Such is the case in, e.g., *45* (E_a = 12 kcal/mole), *47* (E_a = 10 kcal/mole) and *14* for which a very low value of ca. 7 kcal/mole is obtained[89]. Thus in *44* e.g., ring opening by torsion about the 9-8a and 10-1a bonds (see Table 5 for numbering) is always opposed by some of the $H_{4b}-H_{15}$, $H_{4a}-H_{14}$ and $H_{14}-H_{15}$ interactions. In *14* on the other hand the 4 and 5 methyls repel each other for both clockwise and anticlockwise torsions about the two corresponding bonds.

The activation energies and entropies for the thermal ring opening show an isokinetic dependence[11, 23], indicating a common structure of the transition state in all members of the DHP series.

The photochemical ring opening being an allowed process does not usually require any activation energy. In the few exceptions (e.g. *44*, see Table 17) where activation is required the reason seems to be closely connected with the existence of symmetrical steric repulsions mentioned above which stabilize the excited state as well as the ground state[49]. The ring opening is usually the main (if not the only) process taking place in the first excited state of 4a,4b-dihydrophenanthrenes. Its quantum yield ϕ_o is usually about 0.5 or above. In most cases (e.g., *1, 20, 23–25*) the ring opening from the 2nd and 3rd excited states are equally efficient pointing out to the possibility that in these cases the higher singlet states undergo internal conversion to the lowest state. The photochemical ring opening becomes much less efficient in those molecules where it is opposed by steric interactions (e.g., *44*, ϕ_o = 0.08; *46*, ϕ_o = 0.03) and then fluorescence is possible.

The stability of related isomers towards ring opening in both ground and excited state can be compared using electronic overlap populations of the $C_{4a}-C_{4b}$ bonds. Thus in the ground state for *44* n = 0.7274 while for *45* n = 0.7201 reflecting the greater stability of *44* relative to *45*.

VIII Oxidation of 4a,4b-Dihydrophenanthrenes

Due to a unique energetic constellation, the abstraction of the two H atoms from 4a,4b-dihydrophenanthrenes, by molecular oxygen,

$$PH_2 + O_2 \rightarrow P + H_2O_2, * \qquad D'$$

is outstanding, both on account of its chemical mechanism and on account of the importance of quantum mechanical tunnelling in determining the observed reactivity of the initiation step. In the present Section we shall review these topics which were the subject of a series of detailed experimental and theoretical studies of *1*, *20* and of deuterated-*1*[11, 43–47]. These studies were recently extended to cover *44*, deuterated *44*, *41* and deuterated *20*, so that the earlier conclusions are now confirmed by a much wider body of experimental data.

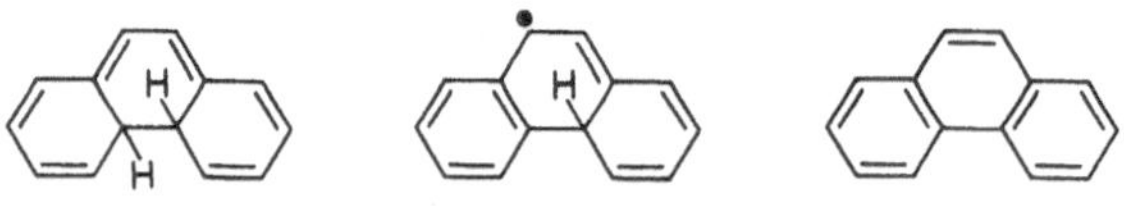

69a, PH_2 *69b*, $PH^{\cdot}$ *69c*, P

Loss of the first H atom from PH_2 gives a free radical $PH^{\cdot}$ which is strongly stabilized due to the formation of one fully aromatic unit. The effective C_{4a}–H bond dissociation energy, D(PH–H) assumes a very low value of ca. 47 Kcal/mole (or below) just because of this aromatic stabilization[44]. This value of D(PH–H) should be compared with the usual C–H bond dissociation energy, e.g., $D((CH_3)_3C{-}H) =$ = 91 Kcal/mole[65].

As a result of an interesting coincidence, the second bond dissociation energy of H_2O_2, $D(H{-}O_2^{\cdot})$ = 47 Kcal/mole[66] happens to be of a similar magnitude as D(PH–H). This coincidence is the reason that in 4a,4b-dihydrophenanthrenes the H atom transfer process

* In this chapter, as in the original papers[11, 43–47], PH_2 and P stand respectively for 4a,4b-dihydrophenanthrene and phenanthrene series molecules. $PH^{\cdot}$ denotes free radical *69b* formed by abstraction of one 4a-H atom from PH_2 and in $PH^{\cdot}$, H denotes one of the "angular" H atoms

$$PH\text{–}H + O_2 \rightarrow PH^{\cdot} + HO_2 \tag{a}$$

for which

$$\Delta H = D(H\text{–}O_2^{\cdot}) - D(PH\text{–}H),$$

is isoenergetic or only weakly exothermic[43–47]. The two consequences to this situation are:

a) Homogeneous, direct, H atom abstraction by O_2

$$R\text{–}H + O_2 \rightarrow R^{\cdot} + HO_2^{\cdot}$$

is an impossible process at ambient temperatures for most hydrocarbons because of its strong endothermicity ($\Delta H \approx 43$ kcal/mole) and therefore cannot serve as an initiation step for hydrocarbon oxidation. Thus in general hydrocarbon oxidation depends for its initiation on catalysts or on an external source of free radicals.

However, in 4a,4b-dihydrophenanthrenes this energetic limiting factor no longer exists, and thus process (a) can (and does) serve as a "built-in" initiation mode. For this reason 4a,4b-dihydrophenanthrenes are oxidized by the "high temperature $HO_2^{\cdot}$ chain" mechanism (see below) even much below room temperature, a path quite inaccessible to other hydrocarbons.

b) Step (a) shows an extremely high kinetic isotope effect of deuterium (e.g., in *1*, $k_a^H/k_a^D \approx 2.5 \times 10^2$ at 221 °K[47]; in *44* $k_a^H/k_a^D > 8 \times 10^2$ at 243 °K[64]). This effect is attributed (in its largest part) to quantum mechanical tunnelling of the free H atom from PH_2 to O_2 across the potential barrier. Such an effect is most prominent for a quasi-isoenergetic reaction such as A in Fig. 8 and smallest for an exothermic process as in B. Going from the initial state at the left to products at the right, the particle (H atom) sees a higher barrier in A than in B[65]. Due to the properties of the barriers the transmission factor in A is larger than in B, where the particle has to tunnel through a wider effective barrier thickness than in A.

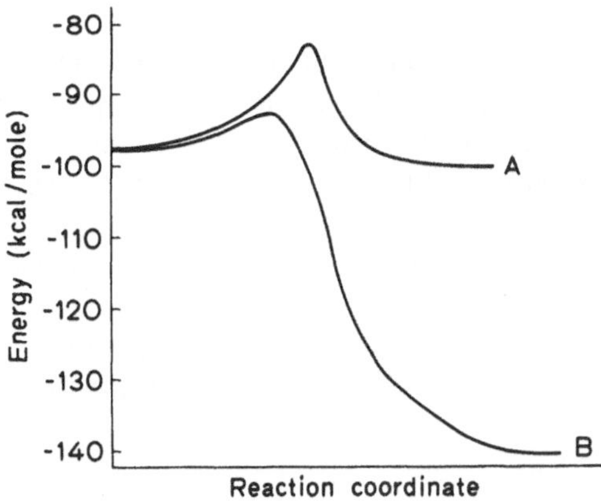

Fig. 8. Energy Profiles for H atom transfer

A Mechanism of the Oxidation of 4a,4b-Dihydrophenanthrenes

All kinetic studies of the oxidation of 4a,4b-dihydrophenanthrenes are facilitated by the possibility of determining their concentration by spectrophotometry in their visible absorption band. This analytical approach has been used in the studies of the oxidation of *1, 20, 41* and *44.*

The oxidation process whose overall stoichiometry is given by D′[11,44] is a chain reaction composed of four distinct primary steps[44]*,

initiation	$PH_2 + O_2 \xrightarrow{k_a} PH^{\cdot} + HO_2^{\cdot}$	(a)
propagation	$PH^{\cdot} + O_2 \xrightarrow{k_b} P + HO_2^{\cdot}$	(b)
	$PH_2 + HO_2^{\cdot} \xrightarrow{k_d} PH^{\cdot} + H_2O_2$	(d)
termination	$2\,HO_2^{\cdot} \xrightarrow{k_e} H_2O_2 + O_2$	(e)

The reaction is susceptible to inhibition, in particular by 2,6-ditert.butyl-substituted phenols *70.*

70a	R=H	*70c*	R=C_2H_5
70b (BHT)	R=CH_3	*70d*	R=n-C_4H_9

Most of the experimental data were obtained with BHT (*70b*). These inhibitors (denoted as SH) compete with PH_2 for the $HO_2^{\cdot}$ radical by processes of type (c), $SH + HO_2^{\cdot} \xrightarrow{k_c}$ inactive products, (c), starting with

t-Bu, t-Bu, OH, R + $HO_2^{\bullet}$ ⟶ tBu, tBu, O•, R + H_2O_2 (c)

(see Ref.[44] for a survey of such processes). Thus when present at sufficiently high concentrations (for *20* and *70b*, $k_d/k_c \approx 10^2$ at 242 °K[44]), inhibitors first stop process (e) and then propagation step (d). When step (d) becomes unimportant, e.g., for $[SH] > 2 \times 10^{-2}$ M, the rate of oxidation of PH_2 is given by the rate of (a),

$$d[PH]_2/dt = -k_a[O_2][PH_2]. \tag{11}$$

Thus by maintaining $[O_2]$ constant (e.g., by having O_2 gas at constant pressure in equilibrium with the solution), the rate of reaction of PH_2 becomes "pseudo-first

* The nomenclature of Ref.[44] is used in the present section

order" in PH_2. This has been indeed shown to be the case over wide ranges of PH_2 concentrations and O_2 pressures.

When inhibitors are present at low concentrations (e.g., $[SH] < 2 \times 10^{-2}$ M), or absent altogether, rather complicated rate laws are obtained for the disappearance of PH_2. Thus starting with the steady state assumption for $PH^{\cdot}$ and for $HO_2^{\cdot}$ the following rate expressions for PH_2 are obtained: in the absence of inhibitors,

$$d[PH]_2/dt = -k_a[O_2][PH_2] - k_d[PH_2]\sqrt{\frac{2k_a}{k_e}[O_2][PH_2]} \tag{12}$$

while at low inhibitor concentrations the expression is

$$d[PH_2]/dt = -k_a[O_2][PH_2] - 2k_a[O_2]\frac{k_d}{k_c}\frac{[PH_2]^2}{[SH]} \tag{13}$$

Depending on the exact conditions, the second right-hand term in (12) is 20–70 times larger than the $k_a[O_2][PH_2]$ term, so that in the absence of inhibitors the rate is given by

$$d[PH_2]/dt \approx -k_d[PH_2]\sqrt{\frac{2k_a}{k_e}[O_2][PH_2]}, \tag{12'}$$

implying a reaction order of 3/2 with respect to $[PH_2]$ and of 1/2 with respect to $[O_2]$. Similarly at $[SH] \approx 10^{-3}$ M equation (13) reduces effectively to

$$d[PH_2]/dt \approx -2k_a[O_2]\frac{k_d}{k_c}\frac{[PH_2]^2}{[SH]}, \tag{13'}$$

indicating a second order reaction with respect to $[PH_2]$, and −1 order with respect to $[SH]$, at constant $[O_2]$.

Expressions (12), (12′), (13) and (13′) and their corresponding integrated forms were applied extensively to the experimental data on the oxidation of *20*[44)], yielding estimates of k_c/k_d and of $k_d/\sqrt{k_e}$.

Considerable use of such expressions was made for obtaining estimates of the kinetic deuterium isotope effect on k_d (e.g., k_d^H/k_d^D) in *1*[46)].

B The Initiation Step of the Oxidation of 4a,4b-Dihydrophenanthrenes[4]

As described in the previous section the presence of inhibitors at high enough concentrations ensures the kinetic "isolation" of the initiation step which becomes the rate determining step in a pseudo first order reaction, the rate of PH_2 disappearance

4 In this section we shall be concerned with the ^1H species. The deuterated molecules will be considered in Section VIII C

being given by Eq. (11). This property simplifies very much the study of the initiation step. Two methods have been used for maintaining the required constant oxygen concentration, depending on the rate constant magnitude. For the more reactive molecules (e.g., *1, 20,* etc.) saturation with O_2 at 1 Atm pressure is adequate ($[O_2]$ in 2,2,4-isooctane is ca. 0.02 M)[47]. For the less reactive molecules, e.g., *41* and *44*, saturation with O_2 at 98.3 Atm ($[O_2]$ in toluene ca. 1 M) allowed working in an easily measurable rate region[64] [5].

The initiation step (as well as the overall oxidation process) can be studied over a wide temperature range, well below ambient temperatures. Thus in *1* it could be studied in the range of 191–263 °K (Table 22)[47], while in *44* the range of 233–303 °K was examined (Table 24)[64]. Apparent Arrhenius activation energies, E_a, Eq. 14,

$$k_a = A_a e^{-E_a/RT} \tag{14}$$

are low, ca. 6–7 Kcal/moles for the ^1H isotopic species (Tables 22–25, the superscripts H and D denoting the isotopic species transferred).

Table 22. Kinetic data for initiation step (a) in *1* and *1*-d_{12} (2,2,4-isooctane solution, 1 Atm O_2, [*70b*] = 0.18 M, k_a^H and k_a^D in $1 \cdot mol^{-1} \cdot h^{-1}$)[a]

T/°K	k_a^H		k_a^D	k_a^H/k_a^D
221	5.5		2.2×10^{-2}	2.5×10^2
242	17.3		0.182	95
263	53.7		0.838	64
E^H = 6.2 Kcal/mol.	A^H = 24,400 ($10^{4.387}$)		$1 \cdot mol^{-1} \cdot s^{-1}$	$\Delta S^{\neq H}$ = −38 eu.[b]
E^D = 9.5 Kcal/mol.	A^D = 23,100 ($10^{4.363}$)		$1 \cdot mol^{-1} \cdot s^{-1}$	$\Delta S^{\neq D}$ = −38 eu.[c]
A^D/A^H = 0.95				

[a] $[O_2]$ = 0.028 M at 221 °K, 0.019 M at 242 °K and 0.017 M at 263 °K [47].
[b] Temperature range 221–263 °K [c] Temperature range 242–263 °K

Table 23. Kinetic data for initiation step (a) in *20* and *20*-d_{10} (k_a^H and k_a^D in $1 \cdot mol^{-1} \cdot h^{-1}$, 2,2,4-isooctane, 1 Atm O_2, [*70b*] = 0.18 M)

T/°K	k_a^H	k_a^D	k_a^H/k_a^D
223[a]	12.1	0.125	97
243	39.4	0.79	50
E_a^H = 6.2 Kcal/mol.	A_a^H = 4270. ($10^{3.630}$)		$\Delta S^{\neq H}$ = −41 eu.[b]
E_a^D = 10.2 Kcal/mol.	A_a^D = 3.72×10^5 ($10^{5.571}$)		$\Delta S^{\neq D}$ = −32 eu.[b]
A_a^D/A_a^H = 87			

[a] For $[O_2]$, see Table 22 [b] Temperature range 223–243 °K

5 One should keep in mind the potential instability of such systems at high oxygen pressures

Table 24. Kinetic data for initiation step (a) in *44* and in *44*-d_2 (Toluene, 98.3 Atm O_2, [*70b*] = 0.18 M, k_a^H and k_a^D in $l \cdot mol \cdot h^{-1}$)[a, 64]

T/°K	k_a^H	k_a^D	k_a^H/k_a^D
233	1.37×10^{-2}	–	–
243	2.31×10^{-2}	2.8×10^{-5}	8×10^2
258	4.67×10^{-2}	4.71×10^{-4}	99
273	0.113	1.34×10^{-3}	84
280.4	0.139	2.5×10^{-3}	56
303	0.337	2×10^{-2}	17
E_a^H = 6.3 Kcal/mol.	$A_a^H = 3.4\ (10^{0.54})\ l \cdot mol^{-1} \cdot s^{-1}$		$\Delta S^{\neq H} = -56$ eu.[b]
E_a^D = 10.7 Kcal/mol.	$A_a^D = 185\ (10^{2.26})\ l \cdot mol^{-1} \cdot s^{-1}$		$\Delta S^{\neq D} = -48$ eu.[c]
A^D/A^H = 53			

[a] $[O_2]$ = 1 M. [b] Temperature range 233–303 °K. [c] Temperature range 258–303 °K.

Table 25. Kinetic data for the initiation step (a) in *41* (Toluene, 98.3 Atm O_2 [*70b*] = 0.18 M)[a 64]

T/°K	233	243	258	280.4
$k_a^H/l \cdot mol^{-1} \cdot h^{-1}$	1.8×10^{-2}	4.7×10^{-2}	0.11	0.25
E^H = 7.1 Kcal/mol,	$A^H = 26.7\ (10^{1.427})\ l \cdot mol^{-1} \cdot s^{-1}$,		$\Delta S^{\neq H} = -51$ eu.[b]	

[a] $[O_2]$ = 1 M [b] Temperature range 233–280.4 °K

As seen in Tables 22–25, the Arrhenius preexponential factors A_a for the initiation step are very low, $10^{4.38}$ in *1*, $10^{3.63}$ in *20*, $10^{1.43}$ in *41* and $10^{0.54}$ in *44*. These are very low values for bimolecular reactions for which values of about 10^{10} are observed and also predicted by the Transition State Theory[69]. Thus step (a) belongs to a class of "slow reactions"[69], some of which might have ionic transition states[69]. The activation entropies $\Delta S^{\neq}$ obtained from the Transition State Theory rate constant expression

$$k_a = kT/h\ e^{\Delta S^{\neq}/R} e^{-E_a/RT} \tag{15}$$

are thus very low too, within the range of –38 eu in *1*, to –56 eu in *44*. These values suggest that the transition state for the initiation step has some fastidious steric requirements both on the mutual PH_2–O_2 orientation as well as on that of the solvent molecules.

The $HO_2^{\cdot}$ product of the initiation step is a polar species. Therefore, it is quite natural to expect the transition state of the initiation step to show some ionic character. That this is so can be seen from the enhancement in k_a in *20* going from 2,2,4-isooctane to ethanol or n-butanol (Table 26).

Part of the low value of A_a is due to the tunnelling effect Γ^*, which has a temperature dependence of the form $\Gamma^* = \Gamma_0^* e^{E_\Gamma/RT}$ (see next section)[47]. Γ_0^{*H} con-

Table 26. Rate constants k_a for initiation step (a) in *20*, in $1 \cdot mol^{-1} \cdot min$, in several solvents[44] [*70b*] = 0.18 M.

T/°K	2,2,4-Isooctane	Ethanol	n-Butanol
242	0.65	1.5	1.7
231	0.19	0.42	0.48

tributes a factor of $10^{-2.3}$ to the observed A^H which in the absence of tunnelling would thus amount to $A = A^H/\Gamma_0^* = 10^{6.7}$, considerably closer to the normal range of 10^{10} [47].

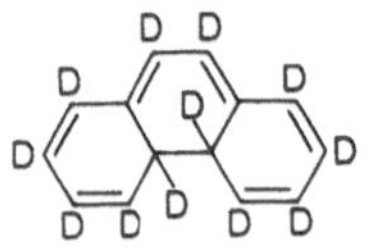

71 (*1*-d$_{12}$)

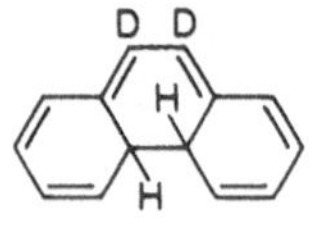

72 (*1*-d$_2$)

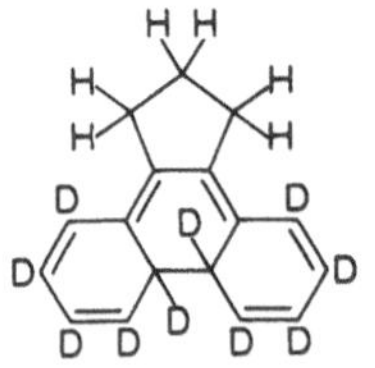

73 (*20*-d$_{10}$)

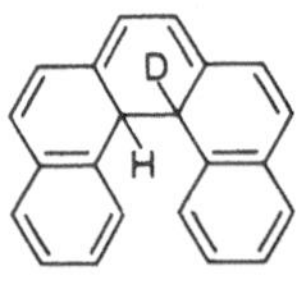

74 (*44*-d)

75 (*44*-d$_2$)

C Kinetic Deuterium Isotope Effects and H Atom Tunnelling in the Oxidation of 4a,4b-Dihydrophenanthrenes

The preliminary work[11] has already indicated the existence of large deuterium kinetic isotope effects in the oxidation of *1*. However, the real extent of the isotope effects could be determined only later, when the role of disproportionation of $HO_2^{\cdot}$ (e) as a termination step, and that of inhibitors in obtaining pseudo first-order rates (Eq. 11) were fully understood[43, 44, 46].

That work has shown that deuterium substitution produced very large kinetic isotope effects on the initiation step (a)[43], and much smaller but still sizable effects on propagation step (d)[46].

Five deuterated species (formulae *71–75*) providing different deuteration patterns in *1, 20* and *44* have been studied up to date[6].

As seen in Tables 22, 23 and 24, very large deuterium kinetic isotope effects k_a^H/k_a^D are common to all 4a and 4b dideuterated 4a,4b-dihydrophenanthrenes. The maximum effects $k_a^H/k_a^D = 2.5 \times 10^2$ in *1*, 10^2 in *20* and $>8 \times 10^2$ in *44* point to a rather uniform behavior which differs in the temperature at which a given k_a^H/k_a^D value is obtained in each molecule. This temperature is highest in *44* and lowest in *20*. Temperatures lower than those in Tables 22–24 would presumably result in still higher isotope effects. The large effect obtained with *44*-d_2 (the largest thus far at 243 °K) clearly shows that the effects in *1*-d_{12} and in *20*-d_{10} are largely *primary* effects. This conclusion was reached previously on the basis of the results obtained with the 4a,4b undeuterated *72* (9, 10-d_2-DHP)[46], in which the secondary effects are small at most, as might have been expected.

Within the sense of the Arrhenius equation, Eq. (14), the isotope effect k_a^H/k_a^D in *1* can be ascribed to a difference $E^D - E^H = 3.3$ Kcal/mol in the activation energies (Table 22) while the preexponential factors are sensibly equal for both isotopic species, $A^D/A^H = 0.95$. This last result, $A^D/A^H \approx 1$ is unusual for reactions showing large tunnelling effects[68], where $A^D/A^H \gg 1$ is usually observed, but is entirely foreseen by the theoretical analysis[46].

For *20*, the isotope effect k_a^H/k_a^D can be traced to a difference $E^D - E^H$ of 4 Kcal/mole in the activation energies which is countered by a preexponential factors ratio of $A^D/A^H = 87$ (Table 23). A similar situation is observed for *44*. Here the Arrhenius activation energy difference amounts to $E^D - E^H = 4.4$ Kcal/mole and the preexponential factors ratio is $A^D/A^H = 53$ (Table 24).

A very large kinetic isotope effect in the initiation stage is also suggested by the experimental results obtained with a 4a monodeuterated DHP[7]. Denoting the rate constant for the abstraction of H atom 4a as k_a' and that for the abstraction of H atom 4b as k_a'' we have

$$k_a = k_a' + k_a'' \tag{16}$$

In a "normal" isotopic species (h_2) we shall have

$$k_a' = k_a'' = 1/2\, k_a. \tag{17}$$

In a 4a monodeuterated molecule (hd) in the presence of a large kinetic isotope effect the 4b ^1H atom will react with a rate constant k_a'' (hd) similar to the rate constant in h_2, e.g.,

$$k_a''(hd) \approx k_a''(h_2) = 1/2\, k_a(h_2), \tag{17'}$$

meaning transferability of rate constants.

---

6 I wish to thank Prof. R. H. Martin (Brussels) for suggesting the synthetic route used for preparing *74* and *75*

7 I am indebted to Prof. D. Arigoni (Zürich) for discussions on this subject

Table 27. 4a-Monodeuteration effect on initiation rate constants k_a (*44*)/k_a(*44*-d). Conditions as in Table 24

T/°K	280.4	273	258	243	233
k_a(*44*)/k_a(*44*-d)	1.97	2.09	2.17	2.13	2.19

For the ^2H atom on 4a we have similarly

$$k'_a(hd) \approx k'_a(d_2) = 1/2\, k_a(d_2), \qquad (17'')$$

(where d_2 denotes the 4a,4b-dideuterated species).

Therefore the initiation rate constant in the monodeuterated molecule

$$k_a(hd) \approx k'_a(d_2) + k''_a(h_2) \approx 1/2\, k_a(h_2) \qquad (17''')$$

will have half its value in the "normal" isotopic species (h_2), as $k_a(d_2) \ll k_a(h_2)$. Table 27 shows that these considerations apply quite well to *44*-d (formula *74*). These results provide an independent check on the mechanism of the oxidation and on the interpretation of the origin of the isotope effect in the initiation step.

Kinetic Isotope Effect in Propagation Step (d)

Step (d), e.g.,

$$PH_2 + HO_2^{\cdot} \xrightarrow{k_d} PH^{\cdot} + H_2O_2 \qquad (d)$$

is strongly exothermic as it involves the cleavage of the weak bond PH–H ($D \approx 47$ Kcal/mole) and the formation of the strong bond HO_2–H ($D = 89$ Kcal/mole[66]). On this account this step is expected to show only limited tunnelling effects and necessarily a much smaller kinetic isotope effect than step (a). Experimentally[46], the kinetic isotope effect on step (d) in *1* was estimated in two ways[46]. In the absence of inhibitors the ratio $k_d/\sqrt{k_e}$ can be evaluated from Eqs. 12 or 12'. In this way, using the known values of k_e^H/k_e^D, the isotopic effect k_d^H/k_d^D was estimated as 7.2 at 263 °K and as 9.9 at 242 °K[46]. At low concentrations of the inhibitor k_d/k_c can be obtained from Eqs. 13 or 13'. In this case k_c can be safely assumed to be independent of the isotopic species, so that this approach can provide a direct estimate of k_d^H/k_d^D. The results obtained by this method are 8.2 at 263 °K and 9.5 at 242 °K.

Theoretical Analysis of Kinetic Isotope Effects in Steps (a) and (d)

Complete theoretical analyses of the kinetic isotope effects have been carried out for steps (a) and (d)[45, 46]. Due to the complexity of the subject only a brief outline will be presented here; the interested reader is referred to Refs. 45 and 46 for full details.

In the theoretical model used the overall rate constant k is represented as the product of the no-tunnelling rate constant k° by the tunnelling correction Γ*,

$$k = k^\circ \cdot \Gamma^* \tag{18}$$

k° is calculated by the Absolute Rate Theory[66, 69], while Γ* is obtained by the procedure of Johnston et al.[70].

In terms of the vibrational partition functions Q of the reactants (′ and ″) and of the activated complex (≠), the Absolute Rate Theory expression for k° is

$$k^\circ = \frac{kT}{h} \cdot \frac{Q^{F}}{Q' \cdot Q''} \tag{19}$$

This expression transforms readily to an Arrhenius type form

$$k^\circ = AT^{-3/2}\, k_v\, e^{-\Delta V_e/RT} \tag{20}$$

where ΔV_e is the electronic energy difference between activated complex and reactants, and A is a temperature independent factor. k_v is a function of the vibrational frequencies ν, of the form

$$k_v = \left\{ \prod_i^{3n-7}{}^{\neq} [1 - \exp(-u_i^{\neq})]^{-1} / \prod_j^{3n-6} [1 - \exp(-u_j')]^{-1} \prod_k^{3n-6} [1 - \exp(-u_k'')]^{-1} \right\} \cdot$$
$$\exp\left[-\frac{1}{2}\left(\sum_i^{3n-7}{}^{\neq} u_i - \sum_j^{3n-6}{}' u_j - \sum_k^{3n-6}{}'' u_k\right)\right], \tag{21}$$

where

$u_i = h\nu_i/kT$, etc.

The tunnelling correction Γ* is the transmission probability through the potential barrier averaged over all possible crossing points and potential energies[69]. An asymmetrical barrier of the Eckart type[70] is assumed in the present model.

The calculations of both k° and of Γ* require a prior determination of the potential energy surfaces which were obtained by the method of Sato[71]. The normal modes calculations were performed by the methods of Warshel et al. (for references cf [45, 46])

Thus, from Eqs. (18) and (20) a kinetic isotope effect k^H/k^D can be expressed as

$$\frac{k^H}{k^D} = \frac{k_v^H}{k_v^D} \cdot \frac{\Gamma^{*H}}{\Gamma^{*D}} \tag{16}$$

where k_v^H/k_v^D is the vibrational contribution, including the zero-point energy loss going from H to D, and Γ^{*H}/Γ^{*D} is the ratio of the tunnelling corrections.

Table 28. Tunnelling and vibrational contributions to kinetic isotope effects in steps (a) and (d)

Step	(a)[45, 47]		(d)[46]	
T	265 °K	242 °K	263 °K	242 °K
Γ^{*H}	118	293	2.70	3.07
Γ^{*D}	14.6	24.8	1.83	2.00
Γ^{*H}/Γ^{*D}	8.1	11.8	1.47	1.53
k_v^H/k_v^D	6.4	7.6	4.0	5.1
k^H/k^D (calc.)	52	90	6.0	7.9
k^H/k^D	64	95	7.2	9.9

Table 28 summarizes some results of the calculations for two temperatures, 242° and 265 °K, for steps (a) and (d) in *1*. The calculated isotope effects, Eq. (20), compare quite well for both steps with the observed effects listed in the last column. This is a very encouraging results as the effects for steps (a) and (d) differ widely in magnitude. The vibrational effects k_v^H/k_v^D for both steps are roughtly comparable while the tunnelling corrections Γ^* and their ratios Γ^{*H}/Γ^{*D} are widely different. The very large tunnelling corrections in step (a), for both H and D atom transfer are thus the direct cause for the large kinetic isotope effect in (a). These tunnelling corrections contribute factors of 8.1 at 263 °K and of 11.8 at 242 °K for step (a), while for step (d) the contributions amount only to 1.47 at 263 °K and 1.53 at 242 °K. The tunnelling corrections for step (d), though much smaller than for (a), are nevertheless not negligible. However, their isotope dependence is insignificant.

How does the present model account for the wide difference in the tunnelling corrections between steps (a) and (d)? The most readily recognizable factors responsible for that difference can be deduced from an examination of the potential surfaces[45–46]. The energy profiles along the reaction coordinates are given in Fig. 8, A for step (a), B for step (d). The electronic activation energies ΔV_e along the reaction path (these are not the effective activation energies) are 15.5 kcal/mole for (a), but only 5.1 kcal/mole for (d). The barrier heights along the –45° sections through the saddle point (ΔV^*) in the potential energy surfaces show the same trend; $\Delta V^* = 2.0$ kcal/mole for step (a) and $\Delta V^* = 0.3$ kcal/mole for step (d). These two parameters (e.g., ΔV_e and ΔV^*) indicate that (a) is much more susceptible than (d) to the effects of tunnelling.

In addition to ΔV_e, both k_v and Γ^* are temperature dependent and thus contribute to the Arrhenius activation energies that may be assigned to Eqs. (18)–(20). The explicit expression which is easily derivable from Eqs. (18)–(20)[45,46] gives for step (a)

$E^H = 6.6$ kcal/mol, $\qquad E^D = 9.8$ kcal/mole

which compare well with the experimental results in Table 22.

Finally it seems appropriate to mention at this stage an obvious conclusion that can be made from the tunnelling effect in Step (a): As suggested recently by Caldin et al.[72] a large tunnelling effect indicates that the transferred H atom does not undergo any significant association with the surrounding solvent molecules.

Acknowledgement. It is great pleasure to express my sincere thanks to my colleagues, past and present members of the Staff of the Photochemical Laboratory for the fine collaboration I enjoyed in these and other topics, in particular to my old-time friend Professor E. Fischer and to my colleagues and associates Drs. D. Gegiou, A. Bromberg, R. Korenstein, Sh. Sharafi-Ozeri, G. Seger, H. Kessel and T. Wismontski. Special thanks are due also to Mr. M. Kaganowitch for his synthetic work and to Mrs. M. Kazes and Mrs. N. Castel for the numerous photochemical measurements.

IX References

1. Lewis, G. N., Magel, T., Lipkin, D.: J. Amer. Chem. Soc. *62,* 2973 (1940)
2. Ciamician, G., Silber, B.: Chem. Ber. *35,* 4128 (1902)
3. Smakula, A.: Z. Phys. Chem. *B25,* 90 (1934)
4. Parker, C. O., Spoerri, P. E.: Nature *166,* 603 (1950)
5. Moore, W. M., Morgan, D. D., Stermitz, F. R.: J. Amer. Chem. Soc. *85,* 829 (1963)
6. a. Hugelshofer, P., Kalvoda, J., Schaffner, K.: Helv. Chim. Acta *43,* 1322 (1960)
   b. Mallory, F. B., Gordon, J. T., Wood, C. S.: J. Amer. Chem. Soc. *85,* 829 (1963)
7. Stermitz, F. R.: Organic Photochemistry, Vol. 1 p 247, O. L. Chapman (ed.), Dekker New York 1967
8. Blackburn, E. V., Timmons, C. J.: Quart. Rev. *123,* 482 (1969)
9. Gilbert, A.: Photochemistry – Specialist Periodical Reports, D. Bryce-Smith (ed.), *3,* 585 (1972); *4,* 667 (1973); *5,* 499 (1974); *6,* 474 (1975); *7,* 371 (1976); *8,* 391 (1977)
10. Martin, R. H.: Angew. Chem. Intern. Ed. *13,* 649 (1974), and references cited therein.
11. Muszkat, K. A., Fischer, E.: J. Chem. Soc., 662 (1967)
12. a. Wismontski-Knittel, T., Muszkat, K. A., Fischer, E.: Mol. Photochem., in press
    b. Watanabe, S., Ichimura, K.: Chem. Lett., 35 (1972)
13. Bromberg, A., Muszkat, K. A., Fischer, E.: Israel J. Chem. *10,* 765 (1972) [M. Frankel Memorial Volume]
14. Muszkat, K. A., Gegiou, D., Fischer, E.: Chem. Comm., 447 (1965)
15. Muszkat, K. A., Kessel, H., Sharafi-Ozeri, S.: Isr. J. Chem. *16,* 291 (1977) [G. Stein Memorial Volume]
16. Doyle, T. D., et al.: J. Amer. Chem. Soc. *92,* 6371 (1970)
17. Naef, R., Fischer, E.: Helv. Chim. Acta *57,* 2224 (1974)
18. Ramey, C. E., Boekelheide, V.: J. Amer. Chem. Soc. *92,* 3681 (1970)
19. Boekelheide, V., Pepperdiene, W.: J. Amer. Chem. Soc. *92,* 3684 (1970)
20. Muszkat, K. A., Sharafi-Ozeri, S.: Chem. Phys. Lett. *42,* 99 (1976)
21. Forster, E. W., Grellmann, K. H., Linschitz, H.: J. Amer. Chem. Soc. *95,* 3108 (1973)
22. Kellog, R. M., Groen, M. B., Wynberg, H.: J. Org. Chem. *32,* 3093 (1967)
23. Blackburn, E. V., Loader, C. E., Timmons, C. J.: J. Chem. Soc., (C) 163 (1970)
24. a. Wismontski, T., Thesis, Ph. D.: The Weizmann Institute of Science, Rehovot (1977)
    b. Wismontski-Knittel, T., Kaganowitch, M., Seger, G., Fischer, E.: Recueil *98,* 114 (1979)
    c. Wismontski-Knittel, T., Fischer, E.: Mol. Photochem. *9,* 67 (1978/79)
    d. Wismontski-Knittel, T., Fischer, E.: J. Chem. Soc. Perkin II, 449 (1979)
25. Wismontski-Knittel, T., Fischer, G., Fischer, E.: J. Chem. Soc. Perkin II, 1930 (1974)
26. Knittel (Wismontski), T., Fischer, G., Fischer, E.: J. Chem. Soc. Chem. Comm., 84 (1972)

27. Wismontski-Knittel, T., Bercovici, T., Fischer, E.: J. Chem. Soc. Chem. Comm., 716 (1974)
28. Korenstein, R., Muszkat, K. A., Fischer, E.: J. Chem. Soc. Perkin II, 564 (1977)
29. Korenstein, R., Muszkat, K. A., Fischer, E.: Helv. Chim. Acta *59*, 1826 (1976)
30. a. Bercovici, T., et al.: Pure and Applied Chem. *24*, 531 (1970)
b. Korenstein, R., Muszkat, K. A., Sharafi-Ozeri, S.: J. Amer. Chem. Soc. *95*, 6177 (1973)
31. Korenstein, R., Muszkat, K. A., Fischer, E.: Helv. Chim. Acta *53*, 2102 (1970)
32. Korenstein, R.: M. Sc. Thesis, Rehovot (1970)
33. Hirshberg, Y.: J. Amer. Chem. Soc. *78*, 2304 (1956)
34. Korenstein, R., Muszkat, K. A., Fischer, E.: Isr. J. Chem. *8*, 273 (1970)
35. Kortüm, G., Bayer, G.: Ber. Bunsenges. *67*, 24 (1963)
36. Kortüm, G., Bayer, G. M.: Angew. Chem. *75*, 96 (1963)
37. Korenstein, R., et al.: J. Chem. Soc. Perkin II, 438 (1976)
38. Korenstein, R., Muszkat, K. A., Fischer, E.: Mol. Photochem. *3*, 379 (1972)
39. Korenstein, R., Muszkat, K. A., Fischer, E.: J. Photochem. *5*, 447 (1976)
40. Korenstein, R., Muszkat, K. A., Fischer, E.: J. Photochem. *5*, 345 (1976)
41. Cuppen, Th. J. H. M., Laarhoven, W. H.: J. Amer. Chem. Soc. *94*, 5914 (1972)
42. Goedicke, Ch., Stegemeyer, H.: Chem. Phys. Lett. *17*, 492 (1972)
43. Bromberg, A., Muszkat, K. A., Fischer, E.: Chem. Comm., 1352 (1968)
44. Bromberg, A., Muszkat, K. A.: J. Amer. Chem. Soc. *91*, 2860 (1969)
45. Warshel, A., Bromberg, A.: J. Chem. Phys. *52*, 1262 (1970)
46. Bromberg, A., Muszkat, K. A., Warshel, A.: J. Chem. Phys. *52*, 5952 (1970)
47. Bromberg, A., et al.: J. Chem. Soc. Perkin II, 588 (1972)
48. Muszkat, K. A., Schmidt, W.: Helv. Chim. Acta *54*, 1195 (1971)
49. Muszkat, K. A., et al.: J. Chem. Soc., Perkin II, 1515 (1975)
50. Sharafi-Ozeri, S.: Ph. D. Thesis, Rehovot, 1976
51. Fischer, G., et al.: J. Chem. Soc., Perkin II, 1569 (1975)
52. Zechmeister, L.: Cis-Trans Isomeric Carotenoids, Vitamins A and Arylpolyenes, Springer, Vienna, 1962
53. Simpson, W. T.: J. Amer. Chem. Soc. *77*, 6164 (1955)
54. Murrell, J. N.: The Theory of the Electronic Spectra of Organic Molecules, Methuen, London, 1963, Ch. 7
55. Forster, L. S.: Theor. Chim. Acta *5*, 81 (1966)
56. For a discussion on the relationship between geometry changes and bond order changes see inter alia: Heilbronner, E., Muszkat, K. A., Schaublin, J.: Helv. Chim. Acta *54*, 58 (1971)
57. Muszkat, K. A.: unpublished
58. Hammond, G. S., et al.: J. Amer. Chem. Soc. *86*, 3197 (1964)
59. Lamola, A. A., Hammond, G. S., Mallory, F. B.: Photochem. Photobiol. *4*, 259 (1965)
60. Sharafi, S., Muszkat, K. A.: J. Amer. Chem. Soc. *93*, 4119 (1971)
61. a. Gegiou, D., Muszkat, K. A., Fischer, E.: J. Am. Chem. Soc. *90*, 12 (1968);
b. Muszkat, K. A., Gegiou, D., Fischer: ibid. *89*, 4814 (1967);
c. Gegiou, D., Muszkat, K. A., Fischer, E.: ibid. *90*, 3907 (1968), and references cited therein
62. Grellmann, K. H., Hentzschel, P., Wismontski-Knittel, T., Fischer, E.: J. Photochem. *11*, 197 (1979)
63. Tinnemans, A. H. A., Laarhoven, W. H., Sharafi-Ozeri, S., Muszkat, K. A.: Recueil *94*, 239 (1975)
64. Muszkat, K. A.: in preparation. Also: 44th Annual Meeting (June 1977), Israel Chemical Society, Abstract PT-3
65. See e.g.: Benson, S. W.: *The Foundations of Chemical Kinetics*, McGraw Hill Book Co., New York, 1960
66. Foner, S. N., Hudson, R. L.: J. Chem. Phys. *36*, 2681 (1962)
67. See e.g.: Bell, R. P.: Chem. Soc. Rev. *3*, 513 (1974); Caldin, E. F.: Chem. Rev. *69*, 135 (1969)
68. Frost, A. A., Pearson, R. G.: Kinetics and Mechanism, John Wiley, New York, 1961, Ch. 5
69. Sharp, T. E., Johnston, H. S.: J. Chem. Phys. *37*, 1541 (1962); Johnston, H. S., Rapp, D.: J. Amer. Chem. Soc. *83*, 1 (1961); Johnston, H. S., Heicklen, J.: J. Phys. Chem. *66*, 532 (1962)

70. Eckart, C.: Phys. Rev. *35*, 1303 (1930)
71. Sato, S.: J. Chem. Phys. *23*, 592, 2465 (1955)
72. Caldin, E. F., Mateo, S.: J. C. S. Faraday I, *71*, 1876 (1975)
73. Jungmann, H., Güsten, H., Schulte-Frohlinde, D.: Chem. Ber. *101*, 2690 (1968)
74. Muszkat, K. A., Sharafi-Ozeri, S.: Chem. Phys. Lett. *20*, 397 (1973)
74. Wood, C. S., Mallory, F. B.: J. Org. Chem. *29*, 3373 (1964)
75. Muszkat, K. A., Sharafi-Ozeri, S.: Chem. Phys. Lett. *20*, 397 (1973)
76. Muszkat, K. A., Seger, G., Sharafi-Ozeri, S.: J. C. S. Faraday II, *71*, 1529 (1975)
77. Geerts-Evrard, F., Martin, R. H.: Private Communication. I thank Prof. R. H. Martin for providing his unpublished experimental results on the photocyclization of these compounds
78. Muszkat, K. A., Sharafi-Ozeri, S.: unpublished results obtained with QCPE program 334[79)]
79. Sharafi-Ozeri, S., Muszkat, K. A.: program CONFI, QCPE 334, Quantum Chemistry Program Exchange, Indiana University, Bloomington 1977
80. Woodward, R. B., Hoffman, R.: The Conservation of Orbital Symmetry, Weinheim: Verlag Chemie, 1969
81. Heilbronner, E., Bock, H.: The HMO Model and its Application. Part 2: Problems with Solutions, New York: Wiley 1976, p. 420
82. Hoffmann, R.: J. Chem. Phys. *39*, 1397 (1963)
83. a. Coulson, C. A.: J. Chim. Phys. *45*, 243 (1948)
    b. Coulson, C. A.: Disc. Faraday Soc. *2*, 9 (1947)
    c. Crawford, V. A., Coulson, C. A.: J. Chem. Soc., 1990 (1948)
84. Scholz, M., Dietz, F., Mühlstadt, M.: Z. Chem. *7*, 329 (1967); Tetrahedron Lett. 665 (1967)
85. Laarhoven, W. H., Cuppen, Th. H. J. M., Nivard, R. J. F.: Rec. Trav. Chim. *87*, 687 (1968)
86. Mulliken, R. S.: J. Chem. Phys. *23*, 1833, 1841 (1955)
87. Muszkat, K. A., Sharafi-Ozeri, S.: Chem. Phys. Lett. *38*, 346 (1976)
88. Sharafi-Ozeri, S., Muszkat, L., Muszkat, K. A.: Z. Naturforsch. *31A*, 781 (1976)
89. Fischer, E.: Private Communication

Received May 17, 1979

Regio- and Stereo-Selectivities in Some Nucleophilic Reactions

Nguyễn Trong Anh

Laboratoire de Chimie Théorique*, Bâtiment 490, F-91405 Orsay, France

Table of Contents

* The Laboratoire de Chimie Théorique is associated with the CNRS (ERA 549).

I Stereochemistry of S_N2 Reactions

To the best of our knowledge, there is no example of a S_N2 reaction with retention of configuration[1] if the reaction center is a saturated carbon atom[2]. However, if the reaction center is a silicon atom, it is possible, by changing the substrate or the nucleophilic reagent, to obtain highly stereoselective reactions with either predominant retention or inversion of configuration. For example, when *1a* and *1b* are

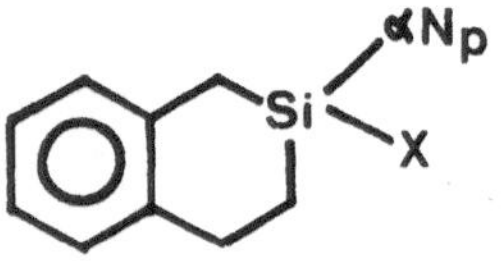

αNp

Si

X

1 a : X = Cl
b : X = OMe
c : X = F

treated with $LiAlH_4$, a substitution reaction occurs with 94% inversion of configuration for the former compound and 96% retention for the latter. Keeping the same leaving group and changing the nucleophile may also induce a reversal of stereoselectivity. Thus, the reaction of *1c* with allyllithium and allylmagnesium bromide occurs with 95% retention and 97% inversion of configuration respectively[3]. It is rather gratifying that all these various stereochemistries may be rationalized with one single perturbational scheme[4].

We start with Salem's treatment of the Walden inversion[5]. Frontier orbital approximation is assumed: the major interaction is supposed to be that between the nucleophile's HOMO and the substrate's LUMO. Now, according to *ab initio* calculations, the latter is essentially an out-of-phase combination of a carbon hybrid atomic orbital ϕ_C with a leaving group hybrid atomic orbital ϕ_X. In the first approximation, the LUMO wave function may be written as:

$$\text{LUMO} \# \phi_C + \frac{\langle \phi_C | P | \phi_X \rangle}{E_C - E_X} \phi_X \tag{1}$$

where E_C and E_X are respectively the energies of ϕ_C and ϕ_X and P is the operator describing the interaction of these two orbitals.

As shown in Fig. 1, the big lobes of these hybrids point toward each other. Therefore, if the nucleophile approaches the substrate from the front side, its HOMO overlaps in phase with the big lobe of ϕ_C and out-of-phase with the big lobe of ϕ_X. Numerical calculations show that the unfavourable (nucleophile-leaving group) interaction usually overrides the favourable (nucleophile – reaction center) interaction in this front-side approach, so that back-side attack is finally preferred, leading to inversion of configuration.

Notice that this back-side attack corresponds to an attack on the small lobe of ϕ_C. It follows that front-side attack may become competitive if it is possible (a) to

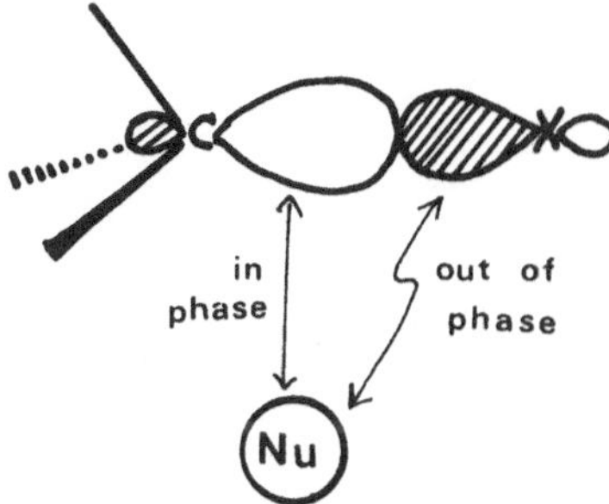

Fig. 1. Overlaps of the nucleophile's HOMO with the substrate's LUMO in a front-side approach leading to retention of configuration

increase the favourable interaction between the nucleophile and the big lobe of the reaction center and (b) to decrease at the same time the unfavourable interaction between the reagent and the leaving group.

One obvious solution is to diminish the mixing coefficient in Eq. (1). This coefficient being inversely proportional to the energy gap $E_C - E_X$, increasing this gap will enhance the contribution of ϕ_C and diminish that of ϕ_X in the LUMO, provided that the integral $\langle \phi_C | P | \phi_X \rangle$ remains substantially the same. This can be done by raising ϕ_C and/or by lowering ϕ_X.

Physically, raising ϕ_C means a decrease of the reaction center's electronegativity. The simplest way to realize this change is by going down the same column in the periodic table. In other words, all things being equal, replacing a carbon atom by a silicon (or germanium or tin or lead) atom as the reaction center will augment the probability of getting retention of configuration. But are "all things" really equal? In fact, the replacement of a carbon atom by a silicon atom introduces also other modifications which, fortunately, favours the same stereochemical trend. Thus, for a given leaving group X, the Si-X bond is longer than the C-X bond and this bond lengthening will put the leaving group farther away from the incoming reagent and reduce their repulsion. At the same time, the valence orbitals of the reaction center become more diffuse. It is therefore possible, at large distances, to have a sizable nucleophile-silicon interaction while the nucleophile-leaving group repulsion remains small: front-side approach is facilitated.

Lowering ϕ_X has less clear-cut stereochemical consequences. When one replaces say X = Cl by X = F, the electronegativity increase and the contraction of the valence orbitals favour retention of configuration but the bond shortening favours inversion of configuration. Numerical calculations suggest that replacing a leaving group by a more electronegative one belonging to the *same column* of the periodic table will increase the percentage of retention of configuration. Indeed, it is known experimentally that F and OR as leaving groups lead to more retention than Cl and SR respectively[6)].

There exists still another way of increasing the favourable (nucleophile-reaction center) interaction in a front-side attack. Consider a tetracoordinated silicon atom *2*. If the R_2SiR_3 angle gets smaller than the tetrahedral value, the R_1SiX angle becomes bigger than 109°28′. This means that the four hybrid atomic orbitals of Si are no longer equivalent: the two used for making the SiR_2 and SiR_3 bonds have less s character than a sp^3 hybrid while the two remaining A.O. acquire more s character. Now

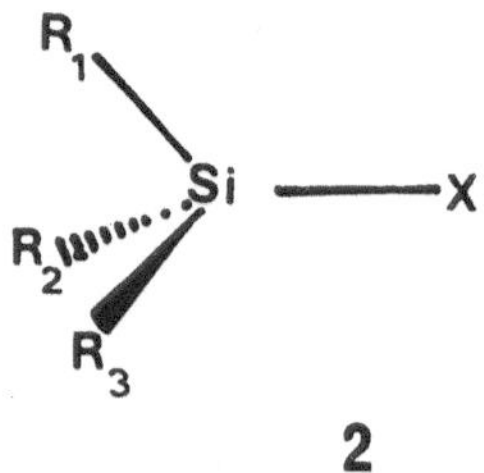

an increase of s character means a greater dissymmetry of the hybrid orbital: the big lobe becomes bigger and the small lobe smaller. Front-side attack on the big lobe is now easier while back-side attack becomes more difficult. It follows that if the Si atom is included in a strained cycle while X remains extracyclic, the percentage of retention of configuration should increase. A similar reasoning shows that if both Si and X are intracyclic, inversion of configuration is favoured. These conclusions agree well with experimental results[7)].

It remains to examine the influence of the nucleophile. Corriu and his co-workers have found an empirical rule according to which "the harder (softer) the nucleophile, the more retention (inversion) of configuration"[8)]. Now a hard reagent is usually a small one, with contracted valence orbitals[9)]. It will overlap little with the leaving group in a front side attack. On the other hand, a soft reagent is usually voluminous, with diffuse valence orbitals[9)]. Its repulsive interaction with the leaving group will be important and the stereochemistry is then shifted toward inversion of configuration.

The foregoing considerations[10)] show that, in order to reproduce the stereochemical trends, it is not necessary to introduce either d orbitals for the silicon atom or pseudo-rotations for the transition state. Conversely, *the stereochemistry of substitution reactions on silicon compounds cannot be taken as a proof of d orbitals intervention in silicon chemistry.*

II The HSAB Treatment of 1,2 vs. 1,4 Additions

The first 1,4 additions of organometallic reagents to conjugated carbonyls were observed as early as 1904[11)]. Half a century later, Eicher still remarked that "at present no unequivocal mechanistic interpretation exists for addition to unsaturated systems" and "further investigations will have to show why in the case of M=Li, (alkyl) transfer leads to 1,2 addition, while in the case of M=Mg, the 1,4-addition mode is preferred; this is certainly not due only to different steric requirements in the (ate) complex or to energetical implications of the transition states involved"[12)]. As a matter of fact, although many experimental results have been gathered between 1904 and 1966, no satisfactory theory has been proposed. Only two empirical rules are known. The first is due to Kohler: an aldehyde usually gives less conjugated addition than an ester or a ketone[13]. The second has been suggested by Gilman and Kirby: the more reactive the organometallic reagent, the more 1,2 addition[14]. No theoretical justification was known for these rules.

An interesting clue is given by the study of regioselectivity in cycloadditions[15]. The cyclodimerization of acrolein gives a single product resulting from the head-to-head orientation:

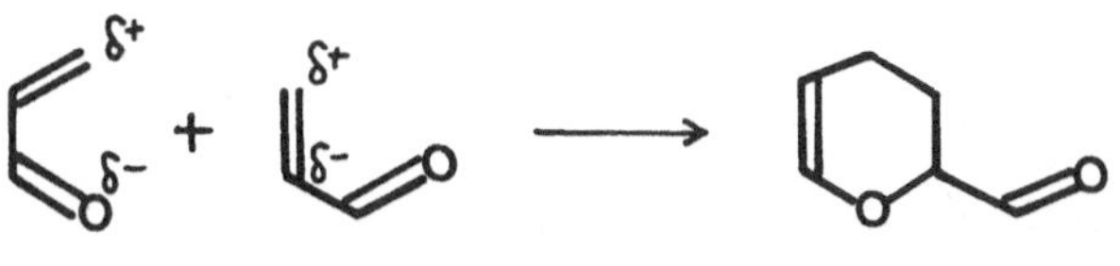

It may be noted that from the electrostatic standpoint, this orientation is most unfavourable as it forces into bonding atoms of like charges. This suggests that attack on C-4 is preferred when the reagent is soft, even if it is positively charged, and *a fortiori* if the reagent is negatively charged. This leads to the rule: a soft (hard) reagent gives preferentially 1,4 (1,2) addition[16].

To determine the hardness (softness) of an organometallic reagent R-M, it is assumed that the hardness of the potential anion R^- parallels that of the cation M^+. Application of Klopman's formula[17] gives the following classification[16]:

Hard: Li^+, Na^+, K^+, Ca^+

Borderline: Al^+, Mg^+

Soft: Cd^+, Hg^+, Ag^+, Cu^+

This classification is in good agreement with experimental trends. It is well known that organoalkalis and organocalcium add preferentially 1,2 to conjugated carbonyls. Organomagnesium and organoaluminium compounds often lead to mixtures of products, while organocadmium and cuprates give predominantly, if not exclusively, conjugated addition. Recently, several groups of organometallic chemists have also used HSAB arguments to rationalize their results[18].

What are the scope and limitations of this HSAB treatment? As Pearson's principle refers to acid-base interactions in general, this treatment is not restricted to organometallic reactions and should apply to other ionic reactions as well. Indeed, it has been successfully extended to reductions of conjugated carbonyls by complex metal hydrides[19] and to Michael-type reactions[20]. Furthermore, it may be shown[16] that Kohler's and Gilman's rules are particular cases of the HSAB rule. There is a slight difference however in the latter case. Gilman and Kirby classify the reagents according to their reactivity. This is a relative property implying a reference compound: a Grignard reagent is reactive with respect to water and unreactive with respect to pentane. On the other hand, the hardness (softness) is an intrinsic property of the molecule and is therefore a more fundamental criterion.

In principle, this constitutes a strong point for the HSAB treatment. In practice, it is its Achille's heel. The reason is that Pearson's rule (a hard reagent attacks preferentially a hard site) should apply only to the reactive species in the transition state.

In other words, to make detailed "predictions", we must already know the reaction mechanism!

Another difficulty exists. It has been found[21)] that in the gas phase or in dipolar aprotic solutions, small anions (cations) have higher HOMO's (lower LUMO's) than the larger one. It follows that interaction between two hard ions is then favoured by both charge control and frontier control. Therefore, the equivalences suggested by Klopman[17)]:

hard + hard = charge-controlled reaction

soft + soft = frontier-controlled reaction

hold only for protic solutions and in the absence of complexing agents (crown ethers, cryptands). For the moment, there is no good substitute for Klopman's theory. However, an important advance has been made in a recent paper by Loupy and Seyden[22)]. Having remarked that in the LUMO of a "free" conjugated ketone, the C-4 coefficient is larger than the C-2 coefficient while the reverse is true for the complexed ketone (compare *3* and *4* in Fig. 2), these authors predicted that conjugated addition

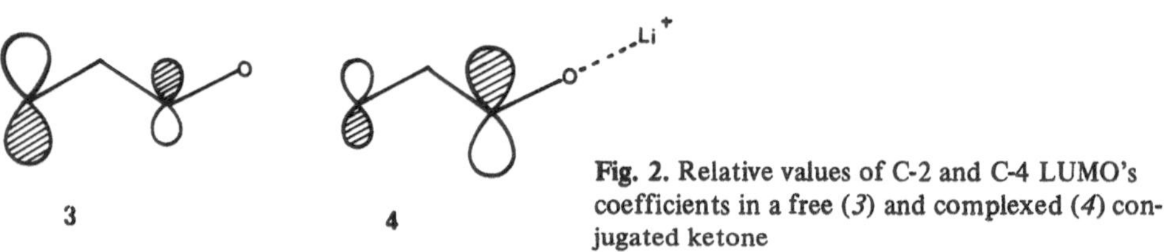

Fig. 2. Relative values of C-2 and C-4 LUMO's coefficients in a free (*3*) and complexed (*4*) conjugated ketone

with a hard reagent is possible, provided that the substrate reacts as a free ketone. Indeed, when cyclohexenone is reduced with an excess of $LiAlH_4$ (ether, 15 mn, room temperature) the regioselectivity is 98% of 1,2 addition and 2% of 1,4 addition for a total yield of 98%. In the presence of the (2.1.1) cryptand, the selectivity is reversed: 77% of 1,4 addition vs. 23% of 1,2 addition (total yield: 80%). Therefore, the regioselectivity is under frontier control in both cases, but in the first experiment, the substrate is complexed by Li^+ while in the second one, it is free.

To summarize, the HSAB principle is a very good first approximation but is usually inadequate for detailed analysis of reaction mechanisms. This is not really surprising. After all, this principle is nothing else than a two parameters relationship: each reactant is characterized by its acidic or basic strength and by its hardness (softness). And obviously, we cannot expect to describe the complexity of chemistry with only two parameters. On the other hand, one should not underestimate its utility. Simple Hückel calculations are also a two parameters treatment where the initial choice of the coulombic and resonance integrals α and β is critical. There is no doubt however that, handled with care, these calculations may give valuable insights. The same may be said for the HSAB principle.

III Additions to Saturated Ketones and 1,2 Asymmetric Induction

The results from the preceding section suggest that the 1,2 vs. 1,4 competition is quite complex and a better understanding of the fundamental processes (role of the counter-ion, of the solvent, factors controlling the addition step . . .) is a necessary prerequisite for a detailed analysis. We therefore turned to the study of nucleophilic additions to chiral saturated carbonyls, using the asymmetric induction as a stereochemical probe.

Ab initio STO-3G calculations[23] were performed on the "supermolecules" formed by a nucleophile and a chiral substrate (2-chloropropanal or 2-methylbutanal). The nucleophile, simulated by H^-, is located at 1.5 Å from the carbon atom, in a direction perpendicular to the carbonyl axis. Hence, to each reaction correspond two diastereoisomeric supermolecules resulting from attack on one face or the other of the π system. These molecules are thus models of the diastereoisomeric transition states. In Fig. III and IV, their relative energies are plotted as a function of their conformations. The solid (dashed) lines correspond to the "transition states" of type *5*

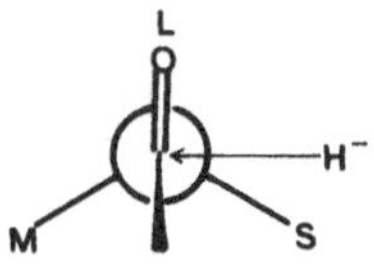

L L
H⁻ H⁻
M S M S

5 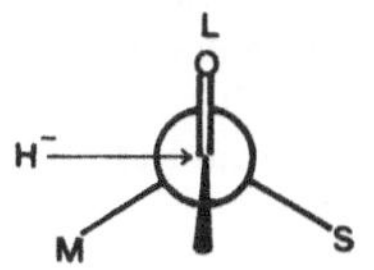 6

(resp. *6*) which will lead to the major (minor) product of the reaction, according to Cram's rule. For 2-chloropropanal, L = Cl, M = Me, S = H and for 2-methylbutanal, L = Et, M = Me and S = H.

If it is assumed that the Curtin-Hammett principle applies, one need only to compare the energies of the minima on the solid and dashed curves to be able to predict the structure of the major product. These curves also allow a direct comparison of Cram's, Cornforth's, Karabatsos's and Felkin's model[24] for 1,2 asymmetric induction. Both Figures show the Felkin transition states lying close to the minima. The Cornforth transition states (Fig. 3) are more than 4 kcal/mol higher and should contribute little to the formation of the final products: assuming a Boltzmann distribution for the transition states, less than one molecule, out of a thousand, goes through them. Similarly, Fig. 4 shows the Cram and Karabatsos transition states to lie more than 2.7 kcal/mol above the Felkin transition states, which means that they account for less than 1% of the total yield.

The foregoing calculations have been done for gas-phase reactions between the naked substrate and the naked nucleophile. This is not a very realistic representation of organic reactions which are usually run in solution and in the presence of the counter-cation. Furthermore, we have assumed a perpendicular attack of the nucleophile on the carbonyl group (cf. *5* and *6*). This is a "classical" hypothesis which is adopted in the four models of asymmetric induction studied, but according to recent works[25], is probably not justified.

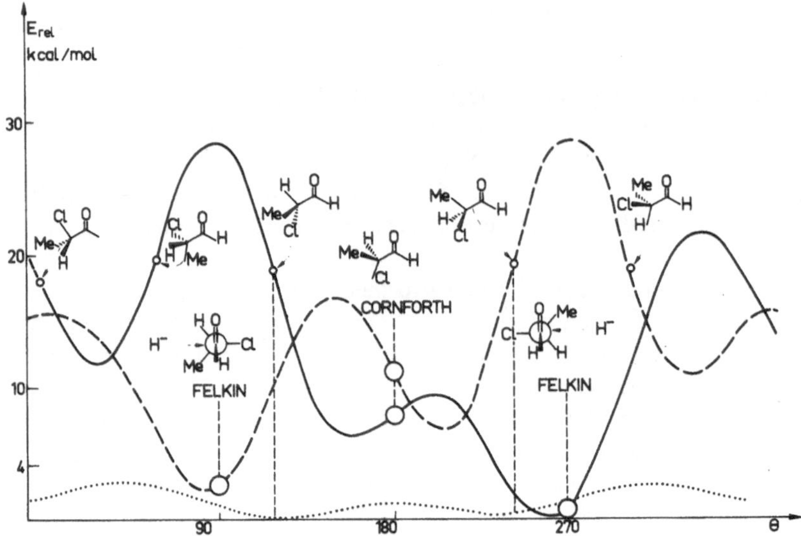

Fig. 3. Reaction of H^- with Me–CHCl–CHO. Plot of transition states energies versus θ, angle of rotation around C1–C2. The solid (dashed) curve corresponds to transitions states *5* (res. *6*) leading to the major (minor) product, according to Cornforth's rule. The dotted curve is the conformational energy curve of Me–CHCl–CHO. (Taken from Ref.[23b)], with permission of the publisher)

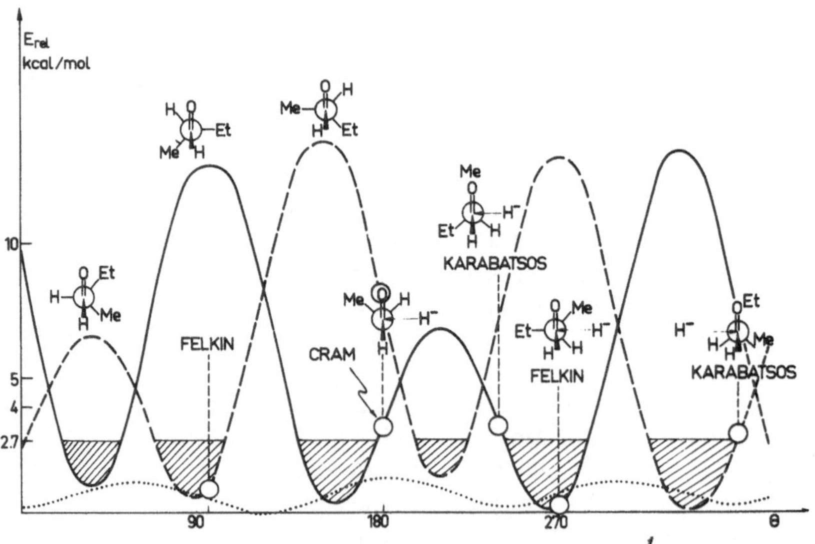

Fig. 4. Reaction of H^- with Et–CHMe–CHO. Same conventions as in Fig. 3 (taken from Ref.[23b], with permission of the publisher)

Calculations were then refined by introducing electrophilic assistance by H^+ or Li^+, solvation of H^- by water[1], and optimization of the angle of nucleophilic attack. These corrections are introduced either separately or simultaneously. It is found[23 b,c)] that the stereochemical influence of anion solvation is negligible, compared with that of the other two factors. In all cases, the shapes of the curves are only slightly modified, as can be seen by comparing Fig. 5 and 6 with Fig. 3 and 4.

As more than 10 series of calculations lead to the same results, it seems reasonable to admit that:

1) It is possible to take into account only a limited number (<6) of conformers of the chiral substrate and thus to avoid the complexities of the Ruch-Ugi[26)] or the Salem[27)] treatments.

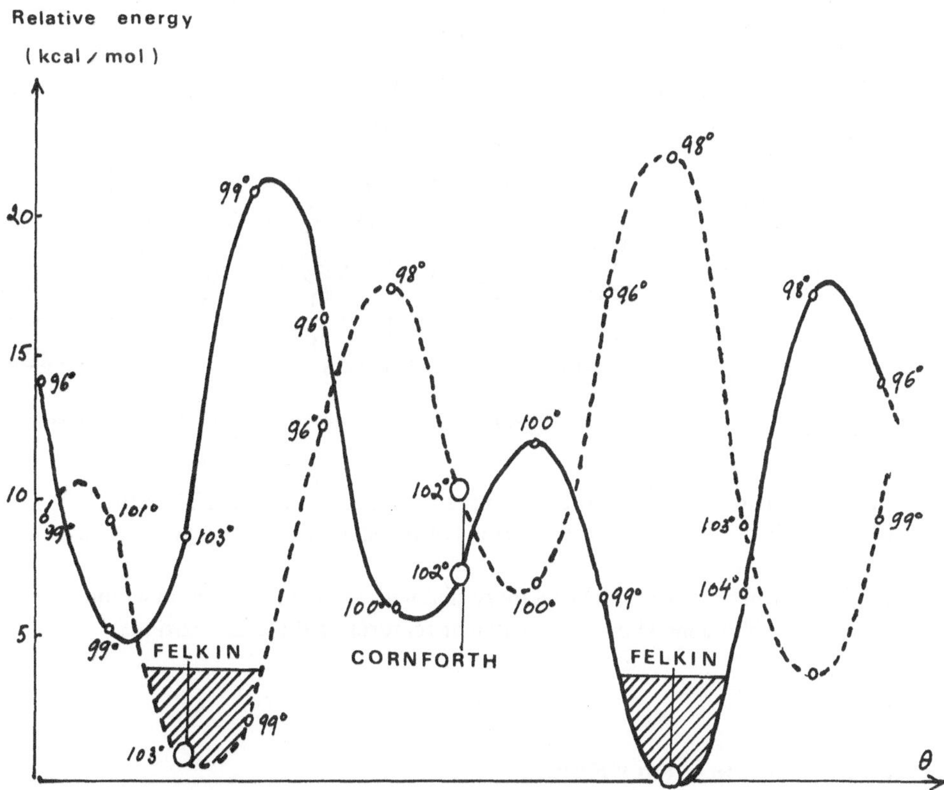

Fig. 5. Reaction of H^- with (Me–CHCl–CHO, Li^+). Angles of nucleophilic attack are optimized

[1] A previous study[21)] has shown that the effects of solvating a nucleophile by a protic or a dipolar aprotic solvent differ in degree and not so much in nature: the major effect remains a lowering of the nucleophile's HOMO level. It is therefore possible, in order to observe the general trend, to solvate the anion with just *one* molecule of water.

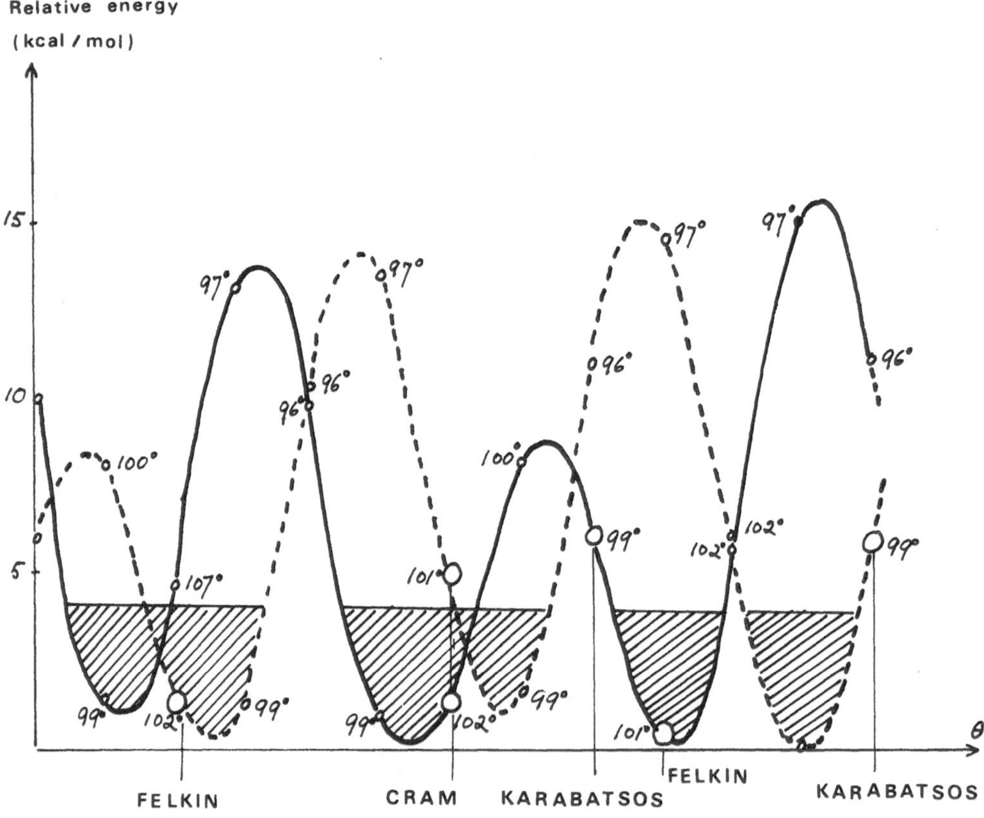

Fig. 6. Reaction of H^- with (Et–CHMe–CHO, Li^+). Angles of nucleophilic attack are optimized

2) Of the four models studied, Felkin's seems to be the best one. The other models, which also lead to correct predictions of the stereochemistry, give rise to rather energetic transition states.

Before we return (in Sect. VI) to a more detailed discussion of Felkin's model, some comments will be made on two particular features of these calculations.

IV How to Violate Cram's Rule

Figures 3, 4, 5, 6 all locate the Cram and Cornforth transition states very near to a crossing of the solid and dashed curves. It follows that a small conformational change of the substrate may reverse the direction of the preferential attack by putting the "dashed" transition state below the "solid" one. The reagent then arrives from the apparently more hindered side, thus violating the Cram's (Cornforth's) rule.

This may be a possible explanation[28] for the remarkable results of Varech and Jacques[29]. In 7, the steric interaction between R and the exo hydrogen tends to

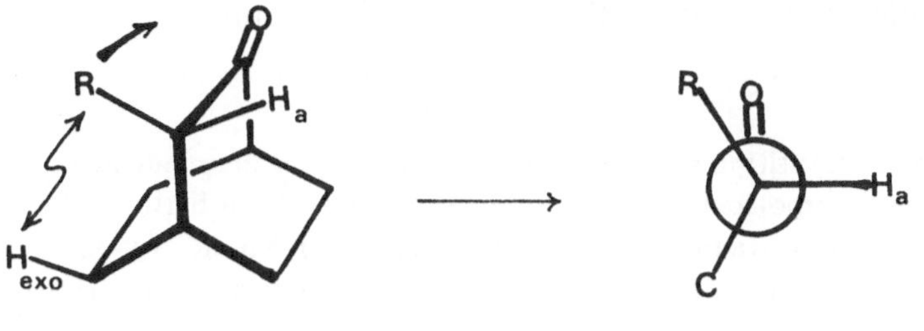

7 8

push the R group upwards. The bridge then takes a Felkin-type conformation *8* and the major product will result from nucleophilic attack *on the side of R.*

In line with this interpretation, it has been experimentally observed that in the reduction of *9* by $LiAlH_4$, preferential attack comes from the left[30]. This can be easily rationalized if one assumes that the methyl substituent tends to move away from the endo-hydrogens:

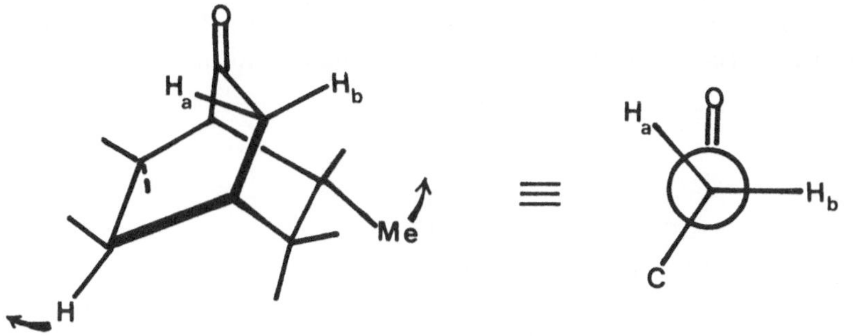

9

Mention should be made however of a completely different interpretation of these results based on the concept of "anisotropic induction effect"[31].

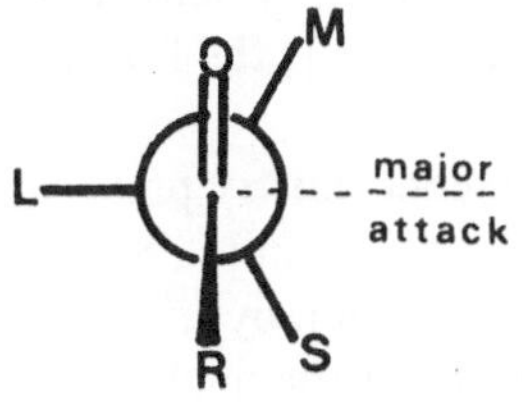

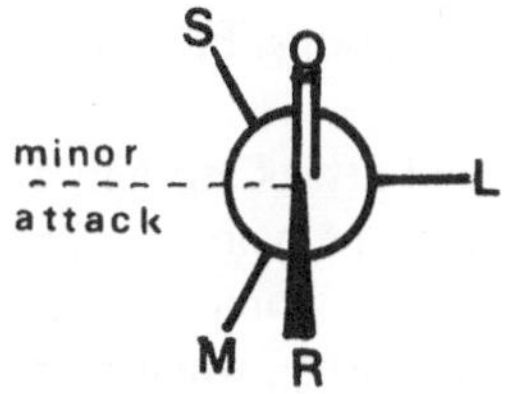

10 11

V The Crucial Role of Non Perpendicular Attack in Asymmetric Induction

Consider the two Felkin transition states *10* and *11*. Obviously, if perpendicular attack is assumed, the intermolecular steric and torsional interactions between the nucleophile and the substrate are identical, all distances (Nu–O, Nu–R, Nu–S, Nu–M) being the same in *10* and *11*. The discrimination can come only from *intra-molecular interactions* in the substrate, which is rather surprising for a bimolecular reaction. In their original paper[24], Felkin and his coworkers postulated that the interactions of substituents M and S are stronger with R than with O. It is not clear why this should be so, as the predominance of R over O must hold even for R = H. Nevertheless, this hypothesis allows the correct prediction of the stereochemistry, agrees well with the observation that selectivity increases with the bulkiness of R, and for these reasons, has not been questioned.

However, Felkin's hypothesis may be advantageously replaced by the assumption of non perpendicular attack. It is clear that the steric hindrance encountered by the nucleophile is much more serious in *13* than in *12*. Furthermore non-perpendicular attack increases the Nu-R interaction at the expense of the Nu-O interaction, thus accounting for the predominance of R over O. Finally, as R becomes bigger, the nucleophile is pushed toward the chiral carbon and can "feel" better the difference between S (in *12*) and M (in *13*), which should lead to an increased selectivity.

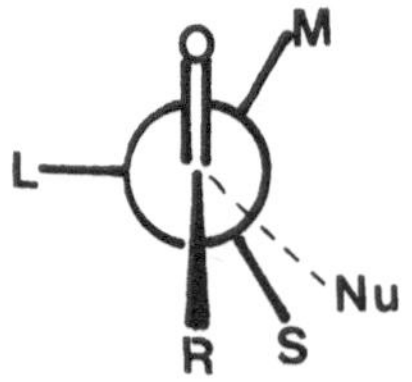

12 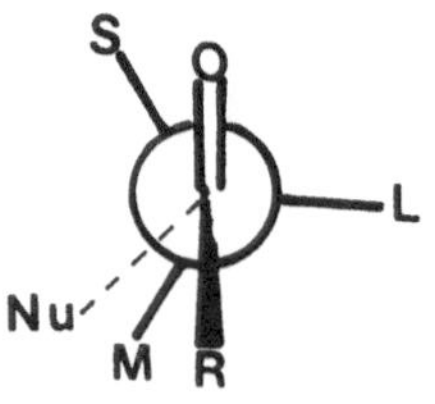 **13**

This qualitative analysis is fully confirmed by STO-3G calculations which supply also some more detailed informations. From the values given in Table 1, the following remarks may be made:

1) Optimizing the angles of attack strongly favours the stereochemistry predicted by Cram's rule: compare the pairs of entries 1 and 2, 3 and 5, 6 and 7, 8 and 9.

2) Electrophilic assistance (i.e. complexation of the carbonyl group by Li^+) favours the "wrong" stereochemistry, when perpendicular attacks of the nucleophile are assumed: compare entries 1 and 3, 6 and 8.

3) Electrophilic assistance reduces the values of the optimal angles of attack (entries 2 and 5, 7 and 9).

Table 1

	Substrate	Reagent	Angle of attack[a]	Asymmetric induction[b]
1	2-chloropropanal	H^-	90°	−1.3 kcal/mol
2	2-chloropropanal	H^-	107° and 103°	−6.95
3	(2-chloropropanal + Li^+)	H^-	90°	0.55
4	(2-chloropropanal + Li^+)	(H^-, H_2O)	90°	0.66
5	(2-chloropropanal + Li^+)	H^-	102°5 and 99°5	−2.05
6	2-methylbutanal	H^-	90°	−0.84
7	2-methylbutanal	H^-	105° and 102°	−4.61
8	(2-methylbutanal + Li^+)	H^-	90°	0.7
9	(2-methylbutanal + Li^+)	H^-	101° and 99°	−0.97

[a] For entries 2, 5, 7 and 9, the first (second) value refers to the optimized angle of attack in transition state *12* (resp. *13*)

[b] As measured by the energy difference between the two Felkin transition states. A negative value corresponds to the stereochemistry predicted by Cram's rule.

4) Solvation of the nucleophile has but little influence on the stereochemistry (entries 3 and 4). Notice however that in our model calculations, only one molecule of solvent is introduced. Therefore, the steric effects of solvation are not adequately accounted for.

From the first three remarks, it follows that the stronger the electrophilic assistance, the smaller the asymmetric induction. Considering that a "hard" nucleophile is generally associated with a highly active cation[22, 32], this observation is strongly reminiscent of the empirical rule suggested by J. Seyden[33]: "soft" nucleophiles give the highest asymmetric induction.

VI Interpretation of Felkin's Model. The Antiperiplanar Effect and the Flattening Rule

It remains to explain why the Felkin transition states are the most stable. During the reaction, the major interaction occurs between the nucleophile's HOMO and the substrate's LUMO. Therefore, the most reactive conformation of the substrate is that with the lowest LUMO. This corresponds to the geometry in which the C2-L bond is parallel to the π system, as there is then a good overlap between the π^*_{CO} orbital and the lowlying $\sigma^*_{C2\text{-}L}$ orbital, leading to a stabilization of the LUMO. The nucleophile may attack this conformer in an antiperiplanar or synperiplanar stereochemistry. The latter is disfavoured for two reasons:

1) while anti attack with respect to L leads to an inphase overlap between H^- and $\sigma^*_{C2\text{-}L}$ (at C2), syn attack leads to an out-of-phase overlap between H^- and $\sigma^*_{C2\text{-}L}$ (Fig. 7)[34].

2) syn attack implies an eclipsing of C1-H^- and C2-L bonds.

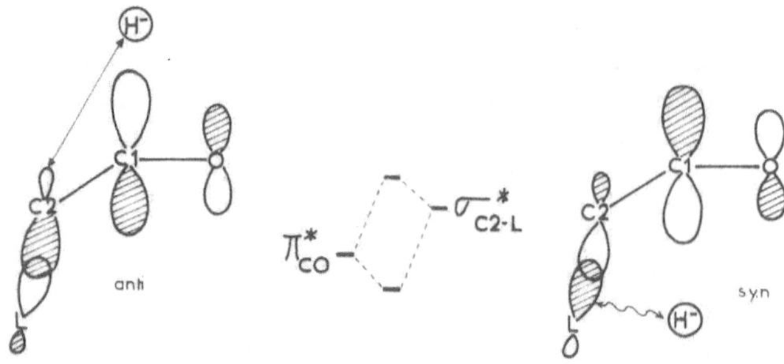

Fig. 7. Secondary interactions for anti- and syn-periplanar attacks (taken from Ref.[23b]), with permission of the publisher)

If this interpretation is correct, then in any reaction with asymmetric induction, a search for antiperiplanarity between the incipient bond (Nu-C1) and an adjacent sigma bond (C2-L) should lead to the most favourable transition states, all other things being equal. Let us apply this rule to the so-called "product development control" problem. Consider a conformationally fixed cyclohexanone, for example *14*.

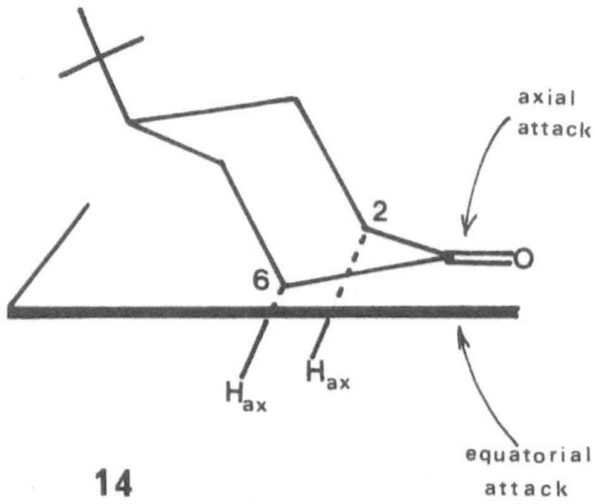

The plane of the carbonyl group defines two half-spaces, the lower one containing only the two axial hydrogens of C2 and C6. It is therefore rather surprising that in many reactions, the nucleophile adds preferentially from the more hindered upper side ("axial attack"). Several explanations have been advanced: product development control[35] torsional repulsions[24e], compression effects[36] . . . Another controlling factor is suggested by the antiperiplanar effect. It is clear that if the ring may be flattened, as indicated below, axial attack may approach antiperiplanarity to the C2-H_{ax} and C6-H_{ax} bonds:

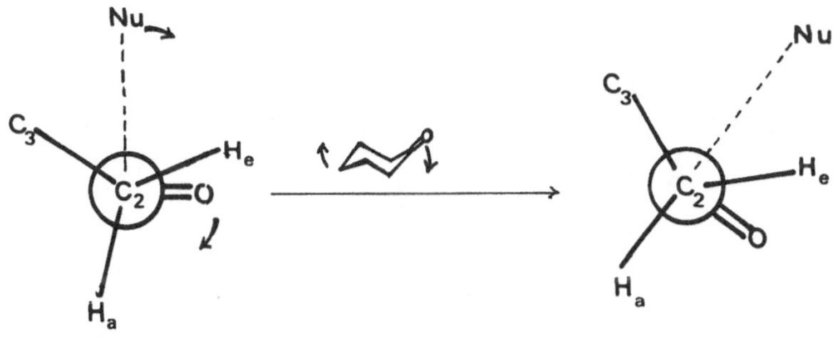

On the other hand, puckering the ring will destroy it: equatorial attack cannot approach antiperiplanarity to the C2–C3 and C5–C6 bonds. We are thus led to the rule[37]: *the more flattened the ring, the more axial attack.* This "flattening rule" may help to rationalize experimental results quite difficult to understand otherwise.

For example, the $LiAlH_4$ reduction of 3-ketosteroids gives 10% β attack while the reduction of 7-ketosteroids gives 55% β attack[38]. The difference does not seem explainable by any known theory. As it appears probable that the steroids B ring is in a chair form, β attack on C7 is an equatorial attack. Torsional repulsions[24e] and compression effect[36] both favour axial attack. Steric hindrance should be at least as important in C7 as in C3 and cannot justify either the increase in β attack. It has been suggested[39] that in fact β attack is as easy at C7 as at C3, but α attack on C7 is hindered by the axial $H_{14\alpha}$. This argument is contradicted by experimental results: the $LiAlH_4$ reduction of *15* – which has the same steric environment as *16* – gives

15 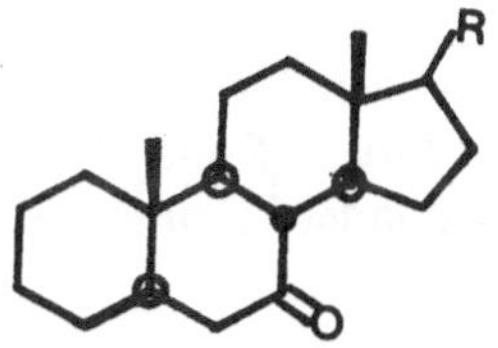 **16**

90% to 95% α attack[40]. The flattening rule provides a simple rationalization: ring B, being linked to two other rings, is less flexible than ring A. Axial (α) attack becomes therefore more difficult on C7 than on C3.

Similarly, ring C is more rigid than ring A. It is also more puckered, being linked by a trans junction to a five-membered ring instead of a six-membered ring[41]. The flattening rule then predicts that axial attack will be easier on 1-ketosteroids than on 12-ketosteroids. This is indeed what is observed by Ayres et al.[42].

Casadevall and Pouet[43] have checked our rule, using bicyclic ketones (*17–20*) of varying flexibilities. Their results are summarized below, along with the analogous results of Suzuki et al.[44]. It is clear that the percentage of axial attack increases with the ketone's flexibility (which may be estimated with the help of Bucourt's dihedral analysis[41]).

17	18	19	20
Percentage of axial attack:			
80% (LAH)	85%	90%	94%
78% ($NaBH_4$)	88%	90%	94%
10% (MeMgI)	32%	43%	55%

21	22	23
60% ($AlMe_3$)	82%	86%

Another verification has been done by Huet and his co-workers[45]. The ketones *24* and *26* are more flattened than *25* and *27* respectively. They should therefore give more axial attack, which indeed is observed.

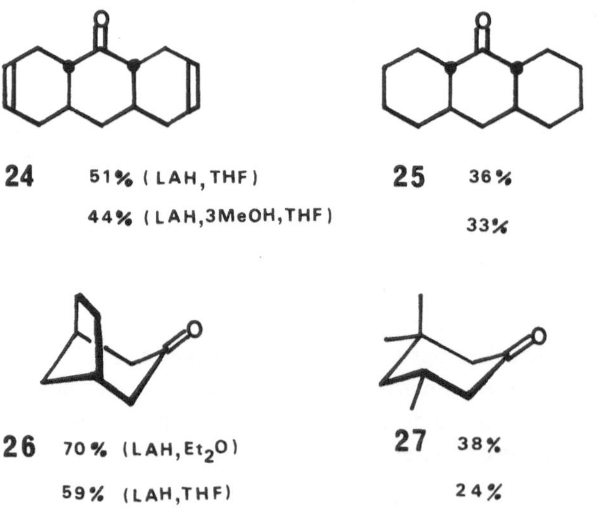

Finally, it should be mentioned that Suzuki et al.[44] have already remarked the correlation existing between ketone flexibility and axial attack. Valls, Toromanoff and Mathieu[46] have also underlined the importance of antiperiplanar attack, although their interpretation is different from ours and stresses the low energy content of the "pre-chair" transition state.

Acknowledgements. My able co-workers, O. Eisenstein, J. M. Lefour and C. Minot are responsible for the essential 95% transpiration part of this work and have also done their share in the 5% inspiration part. Thanks are due to R. Corriu, J. Huet, Y. Maroni, J. Seyden and G. Soussan for many helpful discussions.

VII References

1. Gray, R. W., Chapleo, C. B., Vergnani, T., Dreiding, A. S., Liesner, M., and Seebach, D.: Helv. Chim. Acta *58*, 2524 (1975) and ref. cited
2. Nucleophilic substitutions at vinylic carbon atoms usually proceed with retention of configuration. See, for example: G. Modena, Acc. Chem. Res. *4*, 73 (1971). Rationales have been proposed by: W. D. Stohrer, Tetrahedron Lett., 207 (1975); S. I. Miller, Tetrahedron *33*, 1211 (1977)
3. Massé, J.: Thèse de Doctorat, Montpellier 1969; Corriu, R. and Massé, J.: Chem. Comm., 1373 (1968), J. Organomet. Chem. *35*, 51 (1972)
4. Minot, C.: Thèse de Doctorat, Orsay 1977, Nguyên, Trong Anh and Minot, C., J. Amer. Chem. Soc., in press
5. Salem, L.: Chem. in Britain *5*, 449 (1969). See also: Fukui, K.: Bull. Chem. Soc. Japan *38*, 1749 (1965); Pearson, R. G.: Chem. Eng. News *48*, 66 (1970)
6. Corriu, R. and Henner, B.: J. Organomet. Chem. *102*, 407 (1975)
7. Sommer, L. H., Korte, W. D., and Rodewald, P. G.: J. Amer. Chem. Soc. *89*, 862 (1967); Roark, D. N. and Sommer, L. H.: ibid. *95*, 969 (1973); McKinnie, B. G., Bhacca, N. S., Cartledge, F. K., and Fayssoux, J.: ibid. *96*, 2637 (1974), J. Org. Chem. *41*, 1534 (1976); Dubac, J., Mazerolles, P., and Serres, B.: Tetrahedron *30*, 749, 759 (1974); Cartledge, F. K., Wolcott, J. M., Dubac, J., Mazerolles, P., and Fagoaga, P.: Tetrah. Lett., 3593 (1975); Wolcott, J. M. and Cartledge, F. K.: J. Organomet. Chem. *111*, C35 (1976); Citron, J. D.: ibid. *86*, 359 (1975); Corriu, R. and Massé, J.: ibid. *34*, 221 (1972), Bull. Soc. Chim. Fr., 3491 (1969); Corriu, R., Massé, J., and Guérin, C.: J. Chem. Res. (S) 160 (1977), J. Chem. Res. (M), 1877 (1977); Guérin, C.: Thèse de Doctorat, Montpellier 1978, p. 79
8. Corriu, R. and Lanneau, G.: J. Organomet. Chem. *67*, 243 (1974)
9. Pearson, R. G.: J. Chem. Ed. *45*, 581, 643 (1968)
10. For other theoretical studies of S_N2 reactions, see: Allinger, N. L., Tai, J. C., and Wu, F. T.: J. Amer. Chem. Soc. *92*, 579 (1970); Mulder, J. and Chappell, G.: ibid. *92*, 1819 (1970), Wilhite, D. L. and Spialter, L.: ibid. *95*, 2100 (1973); Lowe, J. P.: ibid. *93*, 301 (1971), *94*, 60 (1972); Epiotis, N. D.: ibid. *95*, 1214 (1973); Dedieu, A. and Veillard, A.: ibid. *94*, 6730 (1972); Keil, F. and Ahlrichs, R.: ibid. *98*, 4737 (1976); Bader, R. F. W., Duke, A. J., and Messer, R. R.: ibid. *95*, 7715 (1973); Van der Lugt, W. and Ros, P.: Chem. Phys. Lett. *4*, 389 (1969); Mulder, J. J. C. and Wright, J. S.: ibid. *5*, 445 (1970); Dyczmons, V. and Kutzelnigg, W.: Theor. Chim. Acta *33*, 239 (1974); Cremaschi, P., Gamba, A., and Simonetta, M.: ibid. *25*, 237 (1972); Schlegel, H. B., Mislow, K., Bernardi, F., and Bottoni, A.: ibid. *44*, 245 (1977); Gillespie, P. D. and Ugi, I.: Angew. Chem. *10*, 503 (1971); Stohrer, W. D.: Chem. Ber. *107*, 1795 (1974), *109*, 285 (1976); Fujimoto, H., Yamabe, S., and Fukui, K.: Tetrah. Lett., 439, 443 (1971); Frenking, G., Kato, H., and Fukui, K.: Bull. Chem. Soc. Japan *49*, 2095 (1976); Baybutt, P.: Mol. Phys. *29*, 389 (1975)
11. Kohler, E. P.: Am. Chem. J. *31*, 642 (1904)
12. Eicher, T.: in "The Chemistry of the carbonyl group", S. Patai Ed., Interscience, London 1966, p. 674

13. Kohler, E. P.: Am. Chem. J. *38*, 511 (1907)
14. Gilman, H. and Kirby, R. H.: J. Amer. Chem. Soc. *63*, 2046 (1941)
15. Eisenstein, O., Lefour, J. M., Nguyên, Trong Anh, and Hudson, R. F.: Tetrahedron *33*, 523 (1977)
16. Eisenstein, O., Lefour, J. M., Minot, C., Nguyên, Trong Anh, and Soussan, G.: C. R. Acad. Sc. *274*, 1310 (1972); Eisenstein, O.: Thèse de 3e cycle, Orsay 1972
17. Klopman, G.: J. Amer. Chem. Soc. *90*, 223 (1968)
18. Inter alia: Priesta, W. and West, R.: J. Amer. Chem. Soc. *98*, 8421 (1976); Seyferth, D., Murphy, G. J., and Mauzé, B.: ibid. *99*, 5317 (1977); Barbot, F., Chan, C. H., and Miginiac, Ph.: Tetrahedron Lett., 2309 (1976)
19. Bottin, J., Eisenstein, O., Minot, C., and Nguyên, Trong Anh: Tetrahedron Lett., 3015 (1972); Durand, J., Nguyên, Trong Anh, and Huet, J.: ibid. 2397 (1974)
20. Deschamps, B., Nguyên, Trong Anh, and Seyden-Penne, J.: Tetrahedron Lett., 527 (1973); Cossentini, M., Deschamps, B., Nguyên, Trong Anh, and Seyden-Penne, J.: Tetrahedron *33*, 409 (1977)
21. Minot, C. and Nguyên, Trong Anh, Tetrahedron Lett., 3905 (1975)
22. Loupy, A., Seyden, J.: Tetrahedron Lett., 2571 (1978)
23. a. Nguyên, Trong Anh, and Eisenstein, O.: Tetrahedron Lett., 155 (1976); b. Nouv. J. Chim. *1*, 61 (1977); c. Eisenstein, O.: Thèse de Doctorat, Orsay 1977
24. a. Cram, D. J. and Elhafez, F. A. Abd: J. Amer. Chem. Soc. *74*, 5828 (1952); b. Cornforth, J. W., Cornforth, R. H, and Mathew, K. K.: J. Chem. Soc. 112 (1959); c. Karabatsos, G. J.: J. Am. Chem. Soc. *89*, 1367 (1967); d. Chérest, M., Felkin, H., and Prudent, N.: Tetrahedron Lett., 2201 (1968); e. Chérest, M. and Felkin, H.: ibid., 2205 (1968)
25. Bürgi, H. B., Dunitz, J. D., Lehn, J. M., and Wipff, G.: Tetrahedron *30*, 1563 (1974) and ref. cited
26. Ruch, E. and Ugi, I.: Top. Stereoch. *4*, 99 (1969)
27. Salem, L.: J. Am. Chem. Soc. *95*, 94 (1973)
28. This interpretation is due to J. M. Lefour
29. Varech, D. and Jacques, J.: Tetrahedron Letters, 4443 (1973)
30. Varech, D.: private communication
31. Chérest, M., Felkin, H., Tacheau, M., Jacques, J., and Varech, D.: Chem. Comm. 372 (1977)
32. Lefour, J. M. and Loupy, A.: Tetrahedron *34*, 2597 (1978)
33. Seyden, J.: Euchem Conference on chirality, La Baule 1972
34. Hoffmann, R.: private communication
35. Dauben, W. G., Fonken, G. J., and Noyce, D. S.: J. Am. Chem. Soc. *78*, 2579 (1956)
36. Schleyer, P. v. R.: J. Am. Chem. Soc. *89*, 701 (1967); Laemmle, J., Ashby, E. C., and Rolling, P. V.: J. Org. Chem. *38*, 2526 (1973)
37. Huet, J., Maroni-Barnaud, Y., Nguyên, Trong Anh, and Seyden-Penne, J.: Tetrahedron Lett., 159 (1976)
38. Fieser, L. F. and Fieser, M.: Steroids, Reinhold, N.Y. 1959, p. 269
39. a. Dauben, W. G., Blanz Jr., E. J., Jui, J., and Micheli, R.: J. Am. Chem. Soc. *78*, 3752 (1956); b. Wheeler, O. H. and Mateos, J. L.: Can. J. Chem. *36*, 1049 (1958)
40. Hückel, W., Maucher, D., Fechtig, O., Kurz, J., Heinzel, M., and Hubele, A.: Liebigs Ann. Chem. *645*, 115 (1961); Grob, C. A. and Tam, S. W.: Helv. Chim. Acta *48*, 1317 (1965); Moritani, I., Nichida, S., and Murakami, M.: J. Am. Chem. Soc. *81*, 3420 (1959)
41. Bucourt, R.: Top. Stereoch. *8*, 159 (1974)
42. Ayres, D. C., Kirk, D. N., and Sawdaye, R.: J. Chem. Soc. (B), 505 (1970)
43. Casadevall, E. and Pouet, Y.: Tetrahedron Letters, 2841 (1976)
44. Suzuki, T., Kobayashi, T., Takegami, Y., and Kawasaki, Y.: Bull. Chem. Soc. Japan *47*, 1971 (1974)
45. Arnaud, C., Accary, A., and Huet, J.: C. R. Acad. Sc. *285*, 325 (1977); Huet, J.: private communication
46. Valls, J. and Toromanoff, E.: Bull. Soc. Chim. France, 758 (1961); Toromanoff, E.: ibid, 708 (1962); Mathieu, J. and Valls, J.: Chem. Weekblad *63*, 21 (1967)

Received June 11, 1979

Author Index Volumes 26–88

The volume numbers are printed in italics

Reactivity and Structure

Concepts in Organic Chemistry

Editors: K. Hafner, J.-M. Lehn, C. W. Rees, P. v. Ragué Schleyer, B. M. Trost, R. Zahradnik

This series will not only deal with problems of the reactivity and structure of organic compounds but also consider synthetical-preparative aspects.
Suggestions as to topics will always be welcome.

Polymers

Properties and Applications

Springer-Verlag
Berlin
Heidelberg
New York

Volume 1
B. Rånby, J. F. Rabek

ESR Spectroscopy in Polymer Research

1977. 356 figures, 29 tables. XIV, 410 pages
ISBN 3-540-08151-8

Contents:

Generation of Free Radicals. – Principles of ESR Spectroscopy. – Experimental Instrumentation of Electron Spin Resonance. – ESR Study of Polymerization Processes. – ESR Study of Degradation Processes in Polymers. – ESR Study of Polymers in Reactive Gases. – ESR Study of the Oxidation of Polymers. – ESR Studies of Molecular Fracture in Polymers. – ESR Studies of Graft Polymerization. – ESR Studies of Crosslinking. – Application of Stable Free Radicals in Polymer Research. – ESR Spectroscopy of Stable Polymer Radicals, Polyradical Anions and their Low Molecular Analogues. – ESR Study of Ion-Exchange Resins. – References. – Index.

Volume 2
H.-H. Kausch

Polymer Fracture

1978. 180 figures, 23 tables. X, 332 pages
ISBN 3-540-08786-9

Contents:

Deformation and Fracture of High Polymers, Definition and Scope of Treatment. – Structure and Deformation. – Statistical, Continuum Mechanical, and Rate Process Theories of Fracture. – Strength of Primary Bonds. – Mechanical Excitation and Scission of a Chain. – Identification of ESR Spectra of Mechanically Formed Free Radicals. – Phenomenology of Free Radical Formation and of Relevant Radical Reactions (Dependence on Strain, Time, and Sample Treatment). – The Role of Chain Scission in Homogeneous Deformation and Fracture. – Molecular Chains in Heterogeneous Fracture.

Volume 3
A. Knop, W. Scheib

Chemistry and Application of Phenolic Resins

1979. 111 figures, 88 tables. XIII, 269 pages
ISBN 3-540-09051-7

Contents:

Historical and Economic Development of Phenolic Resins. – Raw Materials. – Reaction Mechanisms. – Resin Production. – Physiology and Environmental Protection. – Analytical Methods. – Degradation of Phenolic Resins by Heat, Oxygen and High Energy Radiation. – Modified and Thermal-Resistant Resins. – Composite Wood Materials. – Molding Compounds. – Heat and Sound Insulation Materials. – Industrial Laminates and Paper Impregnation. – Coatings. – Foundry Resins. – Abrasive Materials. – Friction Materials. – Phenolic Resins in Rubbers and Adhesives. – Phenolic Antioxidants. – Other Applications. – Index.